DEVELOPMENTS IN IMMUNOLOGY

IMMUNOGLOBULIN GENES
AND B CELL DIFFERENTIATION

Proceedings of the 8th Annual Mid-West Autumn Immunology Conference,
Detroit, Michigan, U.S.A., November 4–6, 1979

Editors:

JACK R. BATTISTO, Ph.D.
Scientific Director, Department of Immunology, Cleveland Clinic Foundation

and

KATHERINE L. KNIGHT, Ph.D.
Professor of Microbiology, University of Illinois Medical Center

ELSEVIER/NORTH-HOLLAND
NEW YORK • AMSTERDAM • OXFORD

Published by:

Elsevier North Holland, Inc.
52 Vanderbilt Avenue, New York, New York 10017

Sole distribution outside of the United States and Canada:

Elsevier/North-Holland Biomedical Press
335 Jan van Galenstraat, P.O. Box 211
Amsterdam, The Netherlands

Library of Congress Cataloging in Publication Data

Mid-west Autumn Immunology Conference, 8th, Detroit, 1979.
    Immunoglobulin genes and B cell differentiation.
    (Developments in immunology; 12 ISSN 0163-5921)

    Bibliography: p.
    Includes index.
    1. Immunology—Congresses. 2. Immunoglobulins—Congresses.
    3. B cells—Congresses. 4. Cell differentiation—Congresses.
    5. Immunogenetics—Congresses. I. Battisto, Jack Richard, 1922–
    II. Knight, Katherine L. III. Title. IV. Series. [DNLM:
    1. Immunoglobulins—Congresses. 2. Genes, Immune response—
    Congresses. 3. B-Lymphocytes—Immunology—Congresses. 4. Cell
    differentiation—Congresses. W1 DE997WM v. 12 / QW 601 M627 1979i]
QR180.3.M52 1979   616.07'9   80-25347
ISBN 0-444-00580-3
    (0-444-80028-X-Series)

Manufactured in the United States of America

Contents

Preface

The Mid-West Autumn Immunology Conference, which is eight years young, has focussed its attention for this year's session upon humoral immunity. Of the two symposia presented, one centered upon how antibody diversity is controlled by immunoglobulin genes and the other dealt with B cell differentiation-activation. Bringing together on one program several experts in each of these topics had the stimulating effect of generating discussion not only on the separate subjects but on the areas of overlap, as well. Of particular interest were the questions of how precursor cells go through the various stages of differentiation to become antibody synthesizing plasma cells, when B cells acquire membrane markers and receptors, how the B cell immunoglobulin surface receptors of antigen function to trigger messages for the cell, when as well as how clonal diversity is brought about, and how the genes that control antibody synthesis are rearranged and RNA is spliced so as to control differences seen in the constant as well as variable regions of the peptide chains that comprise the immunoglobulin molecules.

General discussions at the end of each symposium presentation and at the conclusion of each symposium were designed to foster interaction between the audience and among the symposia speakers. These discussions were taped and appropriately edited versions appear following each speaker's contribution.

The Mid-West Autumn Immunology Conference is also designed to have a number of workshops each of which is presided over by one or two moderators. They are designed to permit participating investigators who have submitted abstracts, to discuss their research interests and problems in small groups. This year the first hour of each workshop session was devoted to a poster session so that participants had the opportunity to examine data more critically and in greater detail. The following two hours were used for short oral presentations of the material appearing in the poster sessions. We have found this format promotes lively discussion by the moderator(s), symposium speakers, and workshop participants. The contents of these workshops have been summarized by the moderator who, depending upon their own inclinations have preceeded the synopses with overviews of the particular areas of research. In this way some moderators have attempted to weave the short individual contributions into the larger fabric of existing knowledge.

Thus, with the exception of the stimulating verbal exchanges that occurred at the Mixer and at the Dinner, the proceedings of the entire Eighth Annual Mid-West Autumn Immunology Conference are contained in this text.

The Editors

Council Members

Mid-West Autumn Immunology Conference
1979

J. R. Battisto, Chairperson
Cleveland Clinic Foundation

W. E. Bullock, Jr.
University of Kentucky

J. E. Butler
University of Iowa

J. F. Clafin
University of Michigan

C. S. David
Mayo Medical School

J. W. Dyminski
Childrens Hospital,
Cincinnati

T. Huard
Postdoctoral Representative
University of Michigan

Y. M. Kong, Treasurer
Wayne State University

H. C. Miller
Michigan State University

H. B. Mullen, Workshop Coordinator
University of Missouri

C. W. Pierce
Washington University

L. S. Rodkey
Kansas State University

J. R. Schmidtke
University of Minnesota

T. Schindler
Graduate Student Representative
University of Illinois

D. Segre
University of Illinois

R. H. Swanborg, Secretary
Wayne State University

J. H. Wallace
University of Louisville

T. G. Wegmann
University of Alberta

Participants

Jerry A. Bash
Georgetown University

Jack R. Battisto
Cleveland Clinic Foundation

Jane Berkelhammer
University of Missouri

Ward E. Bullock, Jr.
University of Kentucky

John Butler
University of Iowa

Max D. Cooper
University of Alabama

David Crouse
University of Nebraska

Bert Del Villano
Cleveland Clinic Foundation

Thomas Feldbush
University of Iowa

James Finke
Cleveland Clinic Foundation

W. Carey Hanly
University of Illinois

Leroy Hood
California Institute of Technology

Norman Klinman
Scripps Clinic and Research

Katherine L. Knight
University of Illinois

Randall Krakauer
Cleveland Clinic Foundation

Stephen P. Lerman
Wayne State University

John Niederhuber
University of Michigan

Dennis Osmond
McGill University

Nicholas Ponzio
Northwestern University

Diego Segre
University of Illinois

Jonathan G. Seidman
National Institutes of Health

Roy S. Sundick
Wayne State University

Ellen S. Vitetta
University of Texas

Randolph Wall
University of California at Los Angeles

Thomas Wegmann
University of Alberta

Acknowledgments

We gratefully acknowledge the generous support of:

NATIONAL INSTITUTE OF ALLERGY AND INFECTIOUS DISEASES
FOGARTY INTERNATIONAL CENTER
ABBOTT LABORATORIES
DIFCO LABORATORIES

We sincerely thank the sustaining sponsors:

ELI LILLY AND COMPANY
MERCK & CO., INC.
THE UPJOHN COMPANY

We also thank the contributing sponsors:

BECKMAN INSTRUMENTS, INC.
MICROBIOLOGICAL ASSOCIATES

I

Immunoglobulin Genes

ORGANIZATION OF IMMUNOGLOBULIN GENES - INTRODUCTORY REMARKS

KATHERINE L. KNIGHT
Department of Microbiology and Immunology, University of Illinois at the
Medical Center, Chicago, Illinois 60680

The genetic origin of antibody diversity has intrigued immunologists and geneticists for many years. Amino acid sequence studies of Bence Jones proteins identified variable (V) and constant (C) regions of immunoglobulin (Ig) light chains and in 1965, Dreyer and Bennett[1] proposed that these variable and constant regions were encoded by separate genes. Strong support for this concept came from continued amino acid sequence studies on mouse and human κ-chains whereby a single constant region sequence was found associated with any one of multiple V region sequences. Likewise, in heavy chains, individual constant regions are associated with any of several V region sequences. In addition, individual V_H regions can be associated with any of the five heavy chain constant regions, C_γ, C_μ, C_α, C_δ, or C_ϵ. For example, allotypes of the variable region of rabbit heavy chains could be found on all classes of Ig molecules.[2,3] The simplest explanation for this observation is that the V_H gene can associate with genes for C_γ, C_μ, C_α, C_δ and C_ϵ. Additional support for the two gene-one polypeptide chain hypothesis came from studies on IgG and IgM monoclonal proteins isolated from one patient;[4,5] idiotypic and amino acid sequence analyses revealed that the V regions of these two molecules were identical whereas the C region represented different Ig classes. Again, the simplest explanation is that V and C regions are encoded by separate genes and that one V_H gene can be associated with both C_γ and C_μ.

Formal proof for the two gene-one polypeptide chain hypothesis was not obtained until 1976 when Tonegawa and his collaborators began direct analysis of the DNA. Initially, they showed that a probe for both the V and C regions of mouse kappa chain (intact kappa chain mRNA) hybridized to two restriction fragments of mouse DNA, whereas a probe for the C region of the kappa chain (the 3'-end half section of the mRNA) hybridized to only one of these two DNA fragments.[6] Thus, the information for V and C regions appeared to be encoded in different DNA fragments. Subsequent studies on the genes coding for mouse lambda chains confirmed and extended these studies.

Embryonic DNA and DNA from a lambda chain plasmacytoma were cleaved by endonucleases and were subjected to agarose gel electrophoresis.[7] The fragments which hybridized to lambda chain mRNA were cloned in a lambda phage vector and the cloned DNA was subjected to R-loop analysis or to nucleotide sequence analysis. By R-loop mapping, the V and C genes were shown to be on different DNA fragments in embryonic DNA and in myeloma DNA, V and C were separated by an intervening sequence of 1250 base pairs.[7] Thus, the V and C regions of lambda chains were also encoded by separate gene segments. Since the V_λ and C_λ genes were much closer together in the myeloma DNA (approximately 1250 base pairs apart) than in the embryonic DNA (the distance between V and C is still unknown) a gene rearrangement must have occurred during differentiation to position the V and C genes closer together, albeit not contiguous. Thus, the Dreyer and Bennett hypothesis of separate genes for V and C had been confirmed.

The excitement over the Ig genes continued. Nucleotide sequence studies showed that in embryonic DNA the codons for the N-terminal 96 amino acids of V_λ were contiguous[8] but the codons for the C-terminal[13] residues of V_λ formed a separate gene segment, designated J_λ; The J_λ gene segment was found between the V_λ and C_λ gene segments, approximately 1250 base pairs to the 5' side of C gene.[9] In lambda chain myeloma DNA, the V_λ and J_λ gene segments were contiguous; thus, the somatic rearrangement which occurred during differentiation resulted in deletion of the intervening sequence between the V and J gene segments. The VJ-C intervening sequence of 1250 base pairs in the myeloma DNA is found in the primary nuclear RNA transcript.[10] This VJ-C intervening sequence is deleted during RNA splicing and the final product is mRNA. The precise nature of the nuclear RNA and the mechanism of RNA splicing are under investigation in Dr. R. Wall's laboratory and will be discussed in detail in his presentation.

The organization of kappa chain genes has been extensively studied by Dr. P. Leder and his collaborators and the essential aspects of the gene organization are similar to those of the lambda chain genes.[11] The variable regions of mouse kappa chains are considerably more heterogeneous than those of lambda chains and examination of this system has allowed an estimate of the number of V_K genes in the germ line.[12] These studies are obviously important to understanding the genetic origin of antibody diversity and Dr. J. Seidman will describe the progress made in this area.

Studies on proteins obtained from patients with heavy chain disease have been of particular interest. Structural analyses of heavy chains isolated from these patients as well as heavy chains of some myeloma proteins have shown non-random deletions.[13,14] Many of the heavy chain mutants have the entire C_H1 domain deleted plus a large portion of the V domain; several other mutants have only the hinge region deleted. In variants where the C_H1 domain was deleted, the deletion usually ended at position 216, the beginning of the hinge region. These observations prompted the suggestion that the constant region of heavy chains may be encoded by more than one gene, one for C_H1, one for the hinge and at least one for the Fc portion of the heavy chain.[13] Thus, the possibility arose that heavy chains may be encoded by at least four genes. Recent studies of embryonic and myeloma DNA have confirmed that indeed, heavy chains are encoded by multiple gene segments, V_H, J, hinge and one for each C_H domain. Studies of the gene organization of heavy chains will be discussed by Dr. L. Hood.

Eventually, the nucleic acid studies can be expected to solve many problems which have arisen from studies of immunoglobulins and their genetic control. For example, such studies should clarify: 1) The origin of antibody diversity, 2) the arrangement of the multiple V_κ and V_H gene segments; 3) the order of the C_H genes; 4) the mechanism of allelic exclusion; 5) the genetic basis for latent allotypes; 6) the mechanism whereby a particular V_H gene is simultaneously expressed with two C_H genes e.g., C_μ and C_δ and 7) how a cell switches from the expression of one C_H gene to another, e.g., C_μ to C_γ or C_α.

REFERENCES

1. Dreyer, W.J. and Bennett, J.C. (1965) Proc. Natl. Acad. Sci., 54, 864.
2. Todd, C.W. (1963) Biochem. Biophys. Res. Commun., 11, 170.
3. Lichter, E.A. (1966) J. Immunol., 98, 139.
4. Nisonoff, A., Fudenberg, H.H., Wilson, S.K., Hopper, J.E. and Wang, A.C. (1972) Fed. Proc., Fed. Am. Soc. Expt. Biol., 31, 206.
5. Wang, A.C., Wang, I.Y.F. and Fudenberg, H.H. (1977) J. Biol. Chem., 252, 7129.
6. Hozumi, N. and Tonegawa, S. (1976) Proc. Natl. Acad. Sci., 73, 3628.
7. Brack, C., Hirama, M., Lenhard-Schuller, R. and Tonegawa, S. (1978) Cell., 15, 1.
8. Tonegawa, S., Maxam, A.M., Tizard, R., Bernard, O., and Gilbert, W. (1978) Proc. Natl. Acad. Sci., 75, 1485.
9. Bernard, O., Hozumi, N. and Tonegawa, S. (1978) Cell., 15, 1133.
10. Gilmore-Herbert, M., Hercules, K., Komaromy, M. and Wall, R. (1978) Proc. Natl. Acad. Sci., 75, 6044.
11. Seidman, J.G., Max, E.E. and Leder, P. (1979) Nature, 280, 370.
12. Seidman, J.G., Leder, A., Edgell, M.H., Polsky, F., Tilghman, S.M., Tiemeier, D.C. and Leder, P. (1978) Proc. Natl. Acad. Sci., 75, 3881.
13. Franklin, E.C. and Fragione, B. (1971) Proc. Natl. Acad. Sci. 68, 187.
14. Fragione, B. (1978) in Comprehensive Immunology: Immunoglobulins, Ed. by Litman, G.W. and Good, R.A., Plenum Med. Book Co., New York, 5, 257.

ORGANIZATION AND REARRANGEMENTS OF HEAVY CHAIN VARIABLE REGION GENES

L. HOOD AND P. EARLY
Division of Biology, California Institute of Technology, Pasadena, California
91125, USA

INTRODUCTION

The advent of recombinant DNA approaches has profoundly expanded our understanding of genes encoding the vertebrate immune response. Serological and
protein chemical analyses had revealed that antibodies were coded for by three
families of genes which lead to the synthesis of light (λ and κ) and heavy (H)
polypeptides. Moreover, the antibody molecule is composed of discrete variable
(antigen recognition) and constant (effector functions) domains. Six extremely
diverse polypeptide segments of the heavy and light chain variable regions,
termed the hypervariable regions, fold to constitute the walls of the antigen-
binding site. Earlier studies revealed that the light chain was encoded by four
separate segments of DNA - leader (L), variable (V), joining (J) and constant
(C)[1,2] (Fig. 1). During differentiation of the antibody-producing or B cells,
the V_L and J_L gene segments are rearranged and joined together to form a
contiguous coding sequence for the variable region (Fig. 1).[1,3,4] The rearranged

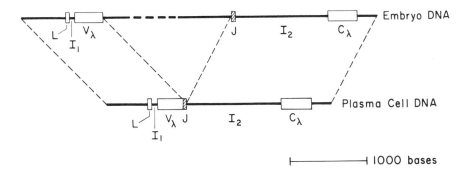

Fig. 1. The organization and rearrangement of mouse λ_I light chain gene segment.
The L exon encodes amino acids present in the precursor to mature light chains.
From reference 1.

or differentiated light chain gene is then transcribed and the remaining intervening nucleic acid sequences are removed by RNA splicing. In this paper we
discuss how the organization of heavy chain gene segments differs from those of

their light chain counterparts and the implications these differences have for mechanisms of DNA rearrangements and antibody diversity.

The organization of heavy chain genes

Since antibody gene segments undergo DNA rearrangements during differentiation, it was important to study heavy chain gene organization in undifferentiated DNAs. We constructed lambda bacteriophage genomic libraries from sperm (undifferentiated) and M603 myeloma (differentiated) DNAs.[5,6] The M603 myeloma tumor synthesizes an IgA immunoglobulin with an α heavy chain. We used a cDNA probe constructed from α chain mRNA to isolate both germline and differentiated antibody gene segments.[5-7]

R-loop mapping and restriction enzyme analyses indicated that the V and C coding regions in the rearranged M603 α gene were separated by 6800 nucleotides[5] (Fig. 2). In addition, each of the coding regions for the three α chain constant domains appears to be separated by an intervening DNA sequence. Thus each of the molecular domains in the antibody molecule appears to have a distinct coding sequence - an observation with important potential evolutionary implications which are discussed elsewhere.[5,7,8]

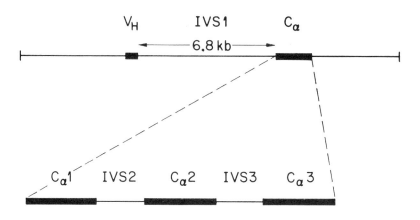

Fig. 2. Structure of the rearranged M603 α gene. Exons encoding each of the four α chain domains (V_H, $C_\alpha 1$, $C_\alpha 2$, $C_\alpha 3$) are separated by intervening DNA sequences.

DNA sequence analyses on the germline V_H and J_H gene segments and on the M603 α gene at the V-J boundary revealed several important features (Fig. 3).[9] First, the germline V_H gene segment terminates at codon 101 - a position homologous to its light chain counterparts. Moreover, the germline J_H gene segment corresponds exactly to the last 15 residues of the V region sequence of the M603 heavy chain. Thus heavy chains are encoded by V_H and J_H gene segments that are rearranged during B-cell differentiation. Second, in the M603 α gene there are five codons lying at the junction between the germline V_H and J_H coding sequences that do not correspond to either of these germline sequences. We have termed this element the D or diversity segment because it constitutes the most highly diverse region of the heavy chain variable region. Because the D segment is not contained within the germline V_H or J_H gene segments, we believe the D segment must be encoded by a third gene segment for heavy chain V regions (i.e., V_H, D and J_H). We have not determined the chromosomal location of the D gene segment, but we believe it must lie between the V_H and J_H gene segments. The D gene segment has important implications for mechanisms of antibody diversity which will be discussed subsequently.

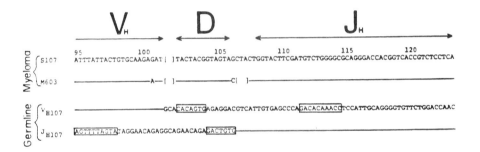

Fig. 3. Comparison of germline V_H and J_H sequences with sequences at the V_H-D-J_H boundaries in rearranged V_H genes. The conserved nucleotides flanking the germline V_H and J_H gene segments are boxed.

Mechanism of V-D-J joining

As suggested by other investigators,[3,4] conserved nucleotides adjacent to germline V and J gene segments appear to be involved in the process of DNA rearrangements to generate the mature variable region gene. We have observed that the same nucleotides occur adjacent to both light and heavy chain gene segments, in each case as blocks of seven and ten conserved nucleotides, separated by a spacer (nonconserved) sequence of 11 or 22 nucleotides (Fig. 4).[9] The conserved sequences adjacent to V gene segments are nearly inverted complements of the sequence adjacent to J gene segments. These observations have led us to propose a model for variable region gene rearrangements, applicable to all three immunoglobulin gene families.[9]

$V_{\kappa 41}$	CACAGTGATACAAATCATAACATAAACC	(11)
$V_{\kappa 2}$	CACAGTGATTCAAGCCATGACATAAACC	(11)
$V_{\kappa 3}$	CACAGTGATTCAAGCCATGACATAAACC	(11)
$V_{\kappa 21}$	CACAGTGCTCAGGGCTGAACAAAAACC	(10)
V_{H107}	CACAGTGAGAGGACGTCATTGTGAGCCCAGACACAAACC	(22)
$V_{\lambda I}$	CACAATGACATGTGTAGATGGGGAAGTAGATCAAGAACA	(22)
$J_{\kappa 1}$	GGTTTTTGTAGAGAGGGGCATGTCATAGTCCTCACTGTG	(22)
$J_{\kappa 2}$	GGTTTTTGTAAAGGGGGGCGCAGTGATATGAATCACTGTG	(23)
*$J_{\kappa 3}$	GGGTTTTGTGGAGGTAAAGTTAAAATAAATCACTGTA	(20)
$J_{\kappa 4}$	AGTTTTTGTATGGGGGTTGAGTGAAGGGACACCAGTGTG	(22)
$J_{\kappa 5}$	GGTTTTTGTACAGCCAGACAGTGGAGTACTACCACTGTG	(22)
J_{H107}	AGTTTAGTATAGGAACAGAGGCAGAACAGAGACTGTG	(21)
J_{H315}	GGTTTTTGTACACCCACTAAAGGGGTCTATGATAGTGTG	(22)
$J_{\lambda I}$	GGTTTTTGCATGAGTCTATATCACAGTG	(11)

Fig. 4. Comparison of noncoding nucleotides adjacent to V and J germline gene segments. Kappa sequences are from references 3 and 4, lambda from reference 1, and heavy chains from reference 9. The mRNA-sense strand is shown 3' to V gene segments and 5' to J gene segments. The asterisk indicates a possibly nonfunctional J gene segment; it has not been found in a rearranged gene. The column to the right lists the spacer length between the underlined conserved hepta- and decanucleotides.

Since the same specific nucleotides and the same spacer lengths are conserved next to all types of variable region gene segments, we propose that these blocks of conserved nucleotides are the recognition sequences for proteins involved in DNA rearrangement. There are apparently two classes of recognition sites: one with an 11-nucleotide spacer between the conserved nucleotides, and one with a 22-nucleotide spacer between the same conserved nucleotides. The complementary recognition sequences adjacent to V and J gene segments are presumably bound by the same protein, but in opposite orientation. Evidently, though, there is a functional difference between proteins which bind to the recognition sequences with 11-nucleotide spacers, and those which bind to sequences with 22-nucleotide spacers, since all DNA joining appears to occur by pairing one gene segment flanked by a recognition sequence with an 11-nucleotide spacer, and one gene segment flanked by a recognition sequence with a 22-nucleotide spacer. Since, for example, all known V_K gene segments have 11-nucleotide spacers and all J_K gene segments have 22-nucleotide spacers, this ensures V_K-J_K joining, rather than V_K-V_K or J_K-J_K. The heavy chain V and J gene segments we have analyzed all have 22-nucleotide spacers and therefore would not be expected to join to one another. As we have suggested, V_H and J_H gene segments probably join with a D gene segment in the middle. The germline D gene segment is presumably flanked on both sides by recognition sequences with 11-nucleotide spacers (Fig. 5).

Fig. 5. Organization of a postulated germline D gene segment. Conserved blocks of 7 and 10 nucleotides with 11-nucleotide spacers would be present on both sides of the D gene segment.

We envision the process of DNA rearrangement in variable region genes as one in which molecules of DNA joining proteins are produced in B-cell precursors and bind to recognition sequences adjacent to V, J, and D gene segments in heavy and light chain immunoglobulin gene families. Probably at low frequencies, proteins bound to recognition sequences with 22-nucleotide spacers encounter proteins bound to recognition sequences with 11-nucleotide spacers, and joining can occur. The intermediate in this reaction would be a structure nearly symmetric about a 2-fold axis with respect to both the conserved nucleotide sequences and the protein molecules bound to each half (Fig. 6). This symmetry is reminiscent of λ repressor bound to one of the phage λ operator sequences.[10]

Gene organization and mechanisms of antibody diversity

An analysis of V_H regions demonstrates that D segment diversity is extensive and variation occurs in size as well as amino acid substitutions (Fig. 7).[11] Moreover, if one analyzes a closely related set of V_H regions, such as those derived from immunoglobulins binding phosphorylcholine,[12] then the third hypervariable region is clearly the most variable portion of the heavy chain.

We believe that the bulk of the variability in the third hypervariable region may be accounted for by three sources.

Germline diversity. The third hypervariable region of the heavy chain includes portions of three distinct gene segments - V_H, D and J_H. Thus germline diversity in each of these elements could contribute to variation in the third hypervariable region. By analogy with light chains, there may be between 200 and

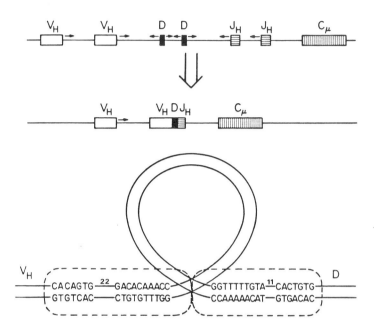

Fig. 6. The upper diagram depicts possible germline D gene segments. Relative distances of the various gene segments from one another are undetermined. The short arrows indicate conserved noncoding sequences (Fig. 4) which may be involved in DNA rearrangement. In this model, DNA rearrangement joins V_H-D-J_H gene segments, or perhaps in some cases just V_H-J_H. Intervening DNA may be deleted, or could undergo other types of rearrangement. The lower diagram shows paired V_H and D (alternatively D and J_H, or V and J) gene segments with 11 and 22 nucleotide spacers. Putative DNA-joining proteins might bind to the areas enclosed by dashed lines. The gene segments are represented as colinear to emphasize the symmetry of the conserved noncoding nucleotides. The actual structure may bring the ends of the two gene segments into close proximity.

500 V_H gene segments.[7] We and others have so far identified four distinct J_H gene segments (9; S. Tonegawa, personal communication). The number of germline D gene segments is as yet unknown.

Combinatorial joining. Antibodies are encoded by split genes whose segments can be rearranged in a combinatorial manner to generate diversity. For example 200 V_H, 10 D and 5 J_H gene segments can be assembled in 10,000 V_H gene combinations (200 x 10 x 5). Thus the combinatorial joining of V, D and J gene segments amplifies antibody diversity enormously. Moreover, the existence of a third gene element (D) in heavy chain genes amplifies the potential diversity generated in V_H regions in an exponential manner as compared to V_L diversity. Thus the presence of the D gene element confers on V_H regions a large amplification in potential diversity over V_L regions.

Protein	Specificity	V segment	D segment	J segment
		75 80 90 99	100 101	102 110 117
M104E	1,3 Dextran	SSSTAYMQLNSLTSEDSAVYYCARD	YD	WYFDVWGAGTTVTVSS
Hdex8	"	---G--------------------	YD	----------------
J558	"	------------------------	RY	----------------
Hdex9	"	--N--F------------------	RY	----------------
Hdex10	"	------------------------	VN	----------------
Hdex6	"	------------------------	SH	----------------
Hdex3	"	------------------------	R-	----------------
Hdex7	"	------------------------	A-	----------------
Hdex2	"	------------------------	NY	----------------
Hdex1	"	------------------------	NY	H----V----------
Hdex5	"	------------------------	SN	Y---Y--Q---L----
Hdex4	"	------------------------	K-	Y---Y--Q---L----
T15,S63,S107,Y5236	Phosphorylcholine	-Q-IL-L-M-A-RA--T-I------	YYGSSY	----------------
H8	"	-Q-IL-L-M-A-RA--T-I------	---N--	----------------
W3207	"	-Q-IL-F-M-A-RA--T-I-----N	--KYDL	--V----------
M603	"	-Q-IL-L-M-A-RA--T-I-----N	----T	-------------
M167	"	-Q-VL-L-M-A-RA--T--T---T--	ADYGDSYF	G------------
M511	"	-Q-IL-L-M-A-RA--T-I------	GDYG-SY	----------------
A4	2,1 Levan	-K-SV-L-M-N-RA--TGI---TTG	[ ]	[ ]-AY--Q--L---
E109	"	-K-SVFL-M-N-RA--TGIH--TTG	[ ]	[ ]-AY--Q--L---
U61	"	-K-SV-L-M-N-RA--TGI---TTG	[ ]	[ ]-AY--Q--L-P-
A47N	"	-K-SV-L-M-N-RA--T-I---STG	[ ]	[ ]-AY--Q--L---
M315	DNP	-ENQFFLK-D-V-[ ]T-T----G-	NDH	L---Y--Q---L---
M21	Unknown	PKN-LFL-MT--R---T-M-----H	GNYPW	YAM-Y--Q--S-----
M173	Unknown	AKN-L-L-MSKVR---T-L-----S	PY	YAM-Y--Q--S-----
T601	Galactan	AKN-L-L-MSKVR---T-L-----G	GYY	G--------------
X44	"	AKN-L-L-MSKVR---T-L-----L	H--	G-AAY--Q--L----A
X24	"	AKN-L-L-MSKVR---T-L-----G	---	G---Y--Q--L----
J539	"	AKNSL-L-MSKVR---T-L-----L	H--	G-NAY--Q--L----A

Fig. 7. A comparison of published heavy chain sequences including the C-terminal portions of V_H regions. The division of this region into V_H segments, D segments, and J_H segments is based on homology with the heavy chains binding dextran. The prototype sequence for the V_H and J_H segments is M104E. The prototypes for the D segments are the first sequence in each group. The J segments which do not extend to amino acid 117 are due to incomplete amino acid sequence data and not to shorter J segments. From reference 11.

V-D-J Junctional diversity. The joining of V, D and J gene segments affords an opportunity for diversity to arise in the junction between the gene segments if the recombinational process is somewhat sloppy. This process is illustrated in Fig. 8 for the joining of a V_K and a J_K gene segment. The recombination between the V and J gene segments can occur at a number of different points and each will generate alternative junctional residues. The theoretical predictions made by this model are precisely born out by the protein sequence data available on appropriate V_K regions.[13]

In heavy chains, variable region gene rearrangement apparently occurs in two steps, V_H-D joining and D-J_H joining. This allows considerable opportunity for junctional variation of the type seen in light chains, as well as mismatch between the reading frames of the V_H and D or D and J_H gene segments. Mismatch in the reading frames of V_H relative to J_H would lead to a nonfunctional immuno-globulin, since the frame shift would carry through to the constant region. Perhaps such abortive rearrangements do occur in many pre-B cells, but these clones would not be selected for expansion. However, the reading frame of the D gene segment itself is not constrained, so long as any frameshift at the V_H-D junction is compensated by a frameshift at the D-J_H junction, and so long as a termination codon does not occur in the D segment. This provides an additional avenue to generate diversity in the third hypervariable region of heavy chains.

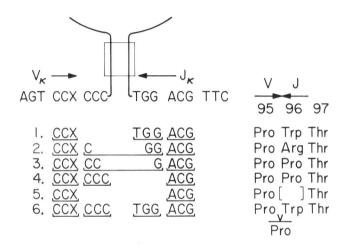

Fig. 8. Junctional diversity of $V_{\kappa 21}$ proteins. Joining occurring at various points between the same germline V_κ and J_κ gene segments generates a series of protein differences at the sites of joining.

Summary

The heavy chain gene organization is distinct from that of its light chain counterparts in that a third gene element, D, encodes the V_H region along with the V_H and J_H gene segments. The V_H and J_H gene segments are flanked on their 3' and 5' sides, respectively, by complementary conserved elements of 7 and 10 nucleotides separated by spacers of 11 or 22 nucleotides. We believe these nucleotides are sites of recognition for joining proteins which mediate V_H, J_H and presumably D gene segment rearrangements and joining. The D gene segment adds in two important regards to the diversification of V_H genes - through its exponential amplification by combinatorial joining and through the generation of two joining sites for mutational recombination.

ACKNOWLEDGEMENTS

This work was supported by an NSF grant.

REFERENCES

1. Bernard, O., Hozumi, N., and Tonegawa, S. (1978) Cell, 15, 1133-1144.
2. Seidman, J. G., Max, E. E., and Leder, P. (1979) Nature, 280, 370-375.
3. Sakano, H., Hüppi, K., Heinrich, G., and Tonegawa, S. (1979) Nature, 280, 288-294.
4. Max, E. E., Seidman, J. G., and Leder, P. (1979) Proc. Nat. Acad. Sci. USA, 76, 3450-3454.
5. Early, P. W., Davis, M. M., Kaback, D. B., Davidson, N., and Hood, L. (1979) Proc. Nat. Acad. Sci. USA, 76, 857-861.
6. Davis, M. M., Calame, K., Early, P. W., Livant, D. L., Joho, R., Weissman, I. L., and Hood, L. (1980) Nature, 283, 733-739.
7. Davis, M., Early, P., Calame, K., Livant, D., and Hood, L. (1979) in Eukaryotic Gene Regulation, Axel, R., Maniatis, T., and Fox, C. F., eds., ICN-UCLA Symposium, Academic Press, New York, pp. 393-406.
8. Calame, K., Rogers, J., Early, P., Davis, M., Livant, D., Wall, R., and Hood, L. (1980) Nature, 284, 452-455.
9. Early, P., Huang, H., Davis, M., Calame, K., and Hood, L. (1980) Cell, 19, 981-992.
10. Ptashne, M., Jeffrey, A., Johnson, A. D., Maurer, R., Meyer, B. J., Pabo, C. O., Roberts, T. M., and Sauer, R. T. (1980) Cell, 19, 1-11.
11. Schilling, J., Clevinger, B., Davie, J. M., and Hood, L. (1980) Nature, 283, 35-40.
12. Hood, L., Loh, E., Hubert, J., Barstad, P., Eaton, B., Early, P., Fuhrman, J., Johnson, N., Kronenberg, M., and Schilling, J. (1976) Cold Spring Harbor Symp. Quant. Biol. 41, 817-836.
13. Weigert, M., Perry, R., Kelley, D., Hunkapiller, T., Schilling, J. (1980) Nature, 283, 497-499.

(See Discussion on following page)

DISCUSSION OF DR. HOOD'S PRESENTATION

Dr. Knight: In the alpha myeloma you showed us, there has been a rearrangement of C_α to the 5' side of C_μ, but is C_μ actually still there?

Dr. Hood: No, C_μ is definitely not there; there are 5000 nucleotides that are to the 5' side of the μ gene. The μ constant gene itself has been rearranged away.

Dr. Knight: But do you know how far? Is it still there or has it been deleted?

Dr. Hood: Well, we cannot answer that question at this time. If the switching mechanism for going from one heavy chain class to another is analogous to the V-J joining mechanism you could postulate that all the intervening DNA would be deleted and the μ gene would be lost. The difficulty in determining whether or not it is lost is that you have the other unrearranged chromosome. If you go into this IgA producing cell and ask if there are gene sequences there, the answer is yes, and they are there in their classic germ line form. We presume these are from the chromosome that did not undergo these rearrangements.

Unidentified person: What do you suppose drives B cells to switch and diversify?

Dr. Hood: The question you are really asking is, given this multi-gene family and given the fact that the information gets read out during development, is the information read out in a linear order programmed fashion? That is, you can envision going down the chromosome 5' to 3', or alternatively the information might be read out in an entirely random fashion. There are phenotypic studies that perhaps Norm Klinman and others will talk about that suggest very strongly that the information may well be read out in a programmed type of fashion. The basic idea that you can begin thinking about is that initial rearrangements set the cell up in such a stage that after it replicates it promotes a next adjacent kind of rearrangement. I think we are going to see that rearrangements at the DNA level are very important, not only for getting V-coding segments and C-coding segments together but also for reorganizing regulatory DNA that programs this information. One of the things that we would all like to do is to define an ideal system in which V genes 1 through 10, for example, are read out in a linear way and then go back to the chromosomes and ask if they are in the same order. It would also be useful to stop the process, possibly through the use of hybridomas, at intermediate stages and to examine the nature of the DNA organization in each of those successive stages. In theory it is possible to contemplate those types of experiments although they are obviously technically extremely difficult.

Dr. Haurowitz: This is very satisfactory for each of us because it makes it much more understandable how, against a single antigen, a thousand different antibodies can be found. It will be very important to find out whether the antigen in any way influences the combination of the many different fragments to form an antibody. According to your experiments, there will be no preformed antibodies as such, but antibodies will be formed from preformed fragments in the presence of the antigen; is this correct?

Dr. Hood: There are two possibilities. The rearrangements that I talked about, at least for the variable coding region, could occur before the organism ever receives antigen. I think that this is the belief of most immunologists. Alternatively, you can't rule out in a totally compeling way the possibility that at least some antibodies and their rearrangements arose after exposure to antigen. It is difficult to understand how the antigen could influence the rearrangements in any informationally useful way, but it remains a logical possibility.

Dr. Knight: Is there evidence for a D region in light chain sequences?

Dr. Hood: No, as far as we can tell both at the protein and the nucleic acid level there is no evidence for a D segment in light chains.

Dr. Wegmann: Do you find V-J joining without D, and V-D joining without J?

Dr. Hood: At the protein level, if you believe the homologies that we have established, there is clear evidence that you could have V to J joining without D. There are no examples of just V and D. You would not expect there to be since it is presumably the J that has the RNA splicing signals, that Dr. Wall will talk about, that allows you to juxtapose the V coding region and the C coding region, at the mRNA level. Moreover, J extends a substantial distance into the third hypervariable region and contributes to diversity. If you deleted that you would certainly modify the molecule in a striking way.

Dr. Knight: From the sperm DNA it appears that all the hypervariable regions are contiguous with the framework residues. This would appear to argue against Dr. Kabats mini-gene theory and the question is how pure was the sperm DNA?

Dr. Hood: The purity of the sperm DNA is greater than 99%. In extensive blot analyses, there is no suggestion whatsoever of any organization other than V segments that are contiguous from one through 99. Overall, I think the probability of the first two hypervariable segments being separate is essentially zero. You must realize, however, that the D gene segment, as Kabat pointed out, is in fact a mini-gene.

Dr. Knight: Yes, so he is certainly correct for at least one of the hypervariable regions.

Dr. Hood: Yes, and one other thing I would say is that in a set of proteins we have for the first time defined the molecular correlations of idiotype. This is a system that has cross reactive idiotypes and individual idiotypes; the cross reactive idiotypes correlate perfectly with several residues in the second hypervariable region and thus you are mapping the V gene segment when you look at the individual idiotype. You can actually take a particular V gene segment and change its V and its J and you can show it retains the same individual idiotype. So, the kinds of things that Dr. Capra talks about with the non-corresponding mapping of idiotypes and hypervariable regions is to a certain extent true as well, in the third hypervariable region.

Dr. Knight: Yes, that is very exciting. Thank you very much Dr. Hood.

DIFFERENT PATTERNS OF IMMUNOGLOBULIN RNA SPLICING

RANDOLPH WALL, EDMUND CHOI, MAUREEN GILMORE-HEBERT[+], MICHAEL KOMAROMY AND JOHN ROGERS

The Molecular Biology Institute and The Department of Microbiology and Immunology, UCLA School of Medicine, Los Angeles, California 90024

INTRODUCTION

Several years ago, we began examining how immunoglobulin mRNAs are made from their nuclear RNA precursors. From our results it is now clear that immunoglobulin mRNAs, like almost all eukaryotic mRNAs studied to date, are generated from much larger nuclear RNA precursors by RNA splicing (reviewed in 1-3). Despite its essential role in higher eukaryotic gene expression, little is known of the signals and molecular events in RNA splicing.

Our approach to resolving these questions regarding RNA splicing involves comparative studies on the immunoglobulin nuclear RNA splicing patterns in normal, immunoglobulin-producing cells and in nonproducing variants which may result from alterations in RNA splicing. Here we summarize our results on the RNA processing pathways which generate different immunoglobulin mRNAs. Our findings indicate that the identical $V \rightarrow C_\kappa$ intervening sequence is spliced out at different times in the processing pathways for two κ mRNAs with the same J-region but different V-regions. We also find that $V \rightarrow C$ splicing in heavy chain mRNA biogenesis occurs in several steps and is therefore more complex than in κ mRNAs. In addition, we propose a mechanism for nuclear RNA splicing in which a specific small nuclear RNA recognizes RNA splicing sites and base pairs with both ends of an intervening sequence so as to align them in exact register for cutting and splicing. Finally, we discuss these findings in relation to the intriguing proposal that immunoglobulin gene expression may be regulated through post-transcriptional RNA splicing.

RNA PRECURSORS AND PROCESSING PATHWAYS FOR IMMUNOGLOBULIN mRNAs.

Initially we constructed and characterized recombinant cDNA clones from MOPC 21 immunoglobulin light and heavy chain mRNAs which provided our first hybridization probes for detecting immunoglobulin nuclear RNA precursors (4, 5). More recently, in collaboration with Dr. Susumu Tonegawa, we have cloned and partially sequenced the MOPC 21 γ_1 heavy chain C-region gene (6). We have ob-

---

[+]present address: The Salk Institute, La Jolla, California.

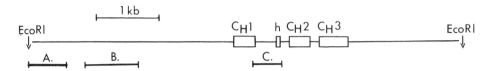

Fig. 1. Hybridization probes from the MOPC 21 heavy chain C-region gene clone. Restriction fragments A (EcoRI/BamHI-595) and B (Pst I-888) represent transcribed intervening sequence probes between the $\gamma 1$ V- and C-regions. A full description of the MOPC 21 clone is presented in Sakano, et al (6). The size of the isolated restriction fragments in base pairs is shown following the flanking restriction sites. Fragment C (Hae III-445) contains the entire 0.33 kb intervening sequence separating the $C_H 1$ domain and the hinge region. The term kb denotes 1000 bases or base pairs.

Fig. 2. Hybridization Probe from the mouse germline κ light chain C-region gene clone. This characterized EcoRI/BamHI segment of the C_κ -J region gene clone was obtained from Dr. Philip Leder (7). Both MOPC 21 and MPC 11 light chains contain J_4 which is located 3.5 kb from the C_κ region. Isolated restriction fragment A (Xba I/Hind III-1000) contains a segment of the J to C_κ intervening sequence.

tained a segment of the C_κ germline gene from Dr. Philip Leder (7). Schematic representations of these clones are shown in Figures 1 and 2. We have now isolated the intervening sequence regions shown and used these to probe the nature and order of removal of intervening sequences in the splicing of specific nuclear RNA processing intermediates.

We previously defined the nuclear RNA processing pathway leading to κ light chain mRNA in MOPC 21 cells (summarized in Figure 3). We obtained formal proof for this κ nuclear RNA processing pathway using glucosamine pulse-chase experiments that showed that all the κ sequences in large nuclear RNA molecules are processed into cytoplasmic mRNA (3). In addition, we determined that the κ C-region transcripts in the nuclear RNAs and in cytoplasmic κ mRNA all showed U.V. target sizes of 10 kb, confirming that they are all derived from a 10 kb transcription unit (3, 8). Polyadenylation of the 10 kb κ primary transcript precedes the RNA splicing steps which generate smaller κ nuclear RNA species (3). Transcript mapping of MOPC 21 κ nuclear RNA species resolved in denaturing CH_3HgOH gels with so-called "Northern blots" (9, 10) using isolated restriction fragment probe A from the MOPC 21 J → C_K intervening sequence shows that this intervening sequence is present in all MOPC 21 κ nuclear RNA species, but not in cytoplasmic κ mRNAs. These transcript mapping results indicate that V → C_K splicing occurs in a single step and is the final processing event in the pathway for MOPC 21 κ light chain mRNAs. This MOPC 21 κ pathway is consistent with 5' → 3' directional RNA splicing events.

We next examined the RNA splicing steps in the MPC 11 κ light chain which has a different V-region, but the same J-region as MOPC 21 (and therefore the same J → C_K intervening sequence in its primary transcript). This system should allow us to determine how this identical intervening sequence is processed in different κ-producing cells. The transcript mapping results summarized in Figure 3 show that the 3.5 kb J-C_K intervening sequence is also removed in a single splicing event, but not as the last step leading to fully processed κ mRNA (11, also Choi, Kuehl and Wall, manuscript in preparation). Accordingly, splicing steps in the MPC 11 κ light chain pathway do not occur in a 5' → 3' orientation.

We have now characterized the γ_1 nuclear RNA species in MOPC 21 cells (Rogers and Wall, manuscript in preparation). The presumptive processing pathway to γ_1 mRNA also involves a number of processing intermediates (Figure 3). We have determined that the U.V. target size of the γ_1 transcription unit through the C_H3 domain is 7-8 kb and therefore in excellent agreement with the largest γ_1 nuclear RNA species. Since all the γ_1 nuclear RNA species contain poly(A), it appears that polyadenylation also precedes RNA splicing in MOPC 21 heavy chain mRNA processing as it does in κ light chain RNA processing.

TABLE 1. RNA SPLICING STEPS IN MOPC 21 HEAVY CHAIN mRNA BIOGENESIS.

| RNA | Hybridization with | | | |
	A.	B.	C.	γ_1-cDNA
7.2 kb	+	+	+	+
5.7	+	+	+	+
4.2	+	o	+	+
2.5	o	o	+	+
2.0 kb	o	o	o	+

See Figure 1 for the origin of these intervening sequence hybridization probes A, B, C. The 2 kb RNA species is γ_1 mRNA.

The results of our heavy chain transcript mapping using isolated intervening sequence probes from the Υ1 clone are presented in Table 1. These data indicate that $V \rightarrow C_{\gamma 1}$ intervening sequences contained in probe fragment B (figure 1) are removed independently of those in probe fragment A. Accordingly MOPC 21 heavy chain γ_1 splicing apparently involves at least two detectable splicing events to join the $V \rightarrow C\gamma_1$-region and does not occur in a 5' $\rightarrow$ 3' direction. The 2.5 kb γ_1 nuclear RNA species closely corresponds to the size of γ_1 mRNA and the intervening sequences separating the $C\gamma_1$ region domains (i.e., 1.97 kb + 0.59 kb = 2.56 kb, ref. 6). This correspondence suggested that the intervening sequences between the different γ_1 C-region domains were removed after γ_1 V $\rightarrow$ C splicing. This prediction appears to be true because the intervening sequence between the $C_H 1$ domain and the hinge region (i.e., probe fragment C, Figure 1) is present in all detectable γ_1 RNA species, but absent from γ_1 mRNA (Table 1).

The processing pathways to these light and heavy chain mRNAs are summarized in Figure 3. Several important conclusions have emerged from these comparative studies on immunoglobulin RNA processing. In all cases examined, polyadenylation precedes RNA splicing. Splicing does not necessarily occur in a 5' $\rightarrow$ 3' direction. The order of RNA splicing steps is apparently determined by factors other than the specific intervening sequence involved. This point is discussed further in the next section which presents a mechanism for RNA splicing. Some intervening sequences are removed in a single event, while others are removed in multiple splicing steps. This suggests that RNA splicing may reveal or generate new splicing substates. Finally, it is clear that potential RNA splicing sites can be somehow ignored. For example, the active MOPC 21 light chain gene in which the V-region is joined to J_4, still contains three J-regions in the resultant $V_{21} \rightarrow C_\kappa$ intervening sequence. These J-regions obviously contain potential splicing sites capable of combining with the C_κ-region. Nonetheless, these potential splicing sites are effectively by-passed in MOPC 21 κ mRNA biogenesis (3).

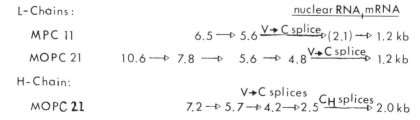

Figure 3. Processing Pathways for Three Immunoglobulin mRNAs. RNA sizes are in kb.

A MECHANISM FOR RNA SPLICING

One clue to the signals possibly used in RNA splicing has emerged from the
nucleotide sequencing of cloned eukaryotic genes which contain intervening sequences. All eukaryotic genes (now including a number of immunoglobulin light
and heavy chain genes with one notable exception) now studied have the characteristic nucleotides; 5'-/GT-intron-AG/-3', at the apparent junctures between
intervening and coding sequences (summarized in 12). We have compiled all
currently available eukaryotic gene sequences for such junctures and have ar-
rived at the following consensus or "optimal" splice sites in which each nucleotide assigned is found in $\geq$45% of all currently known junctures.

$$5'\text{-exon-}\underline{AG}/\underline{GT}AAGTA\text{—intron—TTTT(T)TTTTTTCTTNC}\underline{AG}/G\text{—exon—}3'$$

The upstream and downstream splice sites flanking an intervening sequence
presumably must be brought together for RNA splicing to occur. This could be
achieved by intramolecular basepairing, but no appropriate complementary
structures have been reproducibly found in the vicinity of splice sites. As
an alternative mechanism, we considered that a small effector RNA molecule
might serve to bring splice sites together for RNA splicing. We found that the
most abundant of the stable small nuclear RNAs of eukaryotic cells, U-1 snRNA
(13), is exactly complementary to the "optimal" sequences at RNA splicing sites.
Accordingly, we propose that U-1 snRNA is the recognition component of the
nuclear RNA splicing enzyme, and basepairs with both ends of an intervening
sequence so as to align them in exact register for RNA cutting and splicing
(see Figure 4). There may be an example comparable to this proposed splicing
system in ribonuclease P of E. Coli, which is a site-specific ribonuclease with
an essential RNA component.

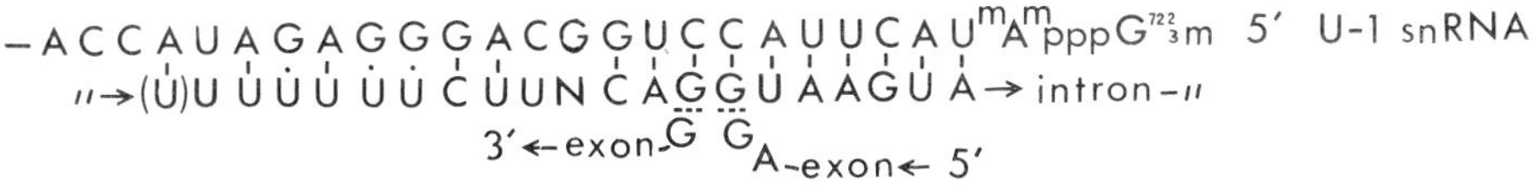

$$-\text{ACCAUAGAGGGACGGUCCAUUCAU}^{m}\overset{m}{A}\text{pppG}^{\prime\prime\prime}\!{}_{3}\text{m} \quad 5' \quad \text{U-1 snRNA}$$

Figure 4. Model for basepairing of U-1 snRNA with Nuclear RNA in the Splicing
Complex. Dots indicate points of cutting and splicing.

A number of properties of U-1 snRNA (reviewed in 14, 15) are consistent with
its being a component of the RNA splicing complex. It is present in large
ribonucleoprotein particles containing hnRNA which are loosely associated with
euchromatin. U-1 RNA is released from the hnRNA isolated from these complexes

by 70% formamide which suggests that it is bound to hnRNA by short regions of basepairing. U-1 is the most abundant snRNA found in cells whose mRNAs are generated by RNA splicing, including human, mouse, chicken and rat. Similar snRNA species also occur in sea urchin, Xenopus and amoeba. U-1 exists in 10^6 copies per cell, and is as stable as ribosomal RNA. Since its overall abundance is the same regardless of the transcriptional activity of the cell type, U-1snRNA is unlikely to be involved in the regulation of eukaryotic gene expression through its availability for RNA splicing.

Our studies on light chain processing pathways indicate a preferred order for RNA splicing steps. The preferential splicing of a particular intervening sequence in a primary transcript containing multiple splice sites could depend on the accuracy of the base pairing of the intron sequences with U-1 snRNA. Alternatively, the splicing pattern of such a transcript may be determined by other factors acting in concert with U-1 snRNA alignment of splicing sites. RNA secondary structure maybe involved. This structure may change as splicing progresses. The geometry of the splicing complex may eliminate some potential sites by placing a limit on the distance between functionally paired upstream and downstream sites. In this regard, it may be significant that no introns less than 66 nucleotides long have been identified in vertebrate systems, although there are many in the 93-140 nucleotide range. The geometry of the splicing complex may require at least 66 nucleotides of intron for the RNA to fold back on itself as indicated in Figure 4. Obviously many other factors may be involved in RNA splicing. Nonetheless, the proposal that U-1snRNA is the recognition component aligning RNA slicing sites provides a starting point for deciphering the molecular events in RNA splicing.

SOME EVIDENCE FAVORING THIS RNA SPLICING MECHANISM

A number of cell lines producing variant immunoglobulin heavy chains having deletions of entire C_H-region domains are known (reviewed in 16). With our discovery that the individual C_H-domains and the hinge in a γ_1 heavy chain gene were all separated by intervening sequences, we proposed that the deletion of entire C_H-domains might result from defects in RNA splicing (6). It now appears that we may have already cloned such a variant γ_1 heavy chain gene. The one reported exception to the /GT...AG/ rule is in the immunoglobulin γ_1 gene from MOPC 21 myeloma cells (6). Recently another copy of the γ_1 gene has been cloned from germline DNA (17), and this conforms to the /GT...AG/ rule. It appears that the trinucleotide GTG following the C_H1 domain is deleted in the MOPC 21 myeloma clone (6). We presume that this MOPC 21 γ_1 gene represents an aberrant gene which would fail to splice at the C_H1-Hinge intron and produce a polypeptide lacking the entire C_H1 domain such as the known γ_1 heavy chain variant IF2 (18). Strikingly, in terms of our proposal on RNA splicing, the MOPC 21 γ_1 C_H1/intron sequence lacking GTG (6) cannot effectively form the core of contiguous base pairs with U-1snRNA to align ends of the C_H1-hinge intervening sequence for splicing. Other such heavy chain deletion variants now being examined both with regard to RNA processing and DNA sequences should serve to further define the signal sequences and molecular mechanisms involved in RNA splicing.

As an aside,another apparent exception to the /GT...AG/ rule is in the third J-region in the mouse κ light chain gene J-region cluster (7, 19). The sequence immediately following this J-region is identical to those following the other four κ J-regions and to the "optimal" upstream intervening sequence juncture except that /GT...is replaced by /CT... Failure to react with U-1 snRNA and splice at this point may explain why this is the only J-region not expressed in any known immunoglobulin κ light chains (7).

CLOSING REMARKS: THE REGULATION OF IMMUNOGLOBULIN GENE EXPRESSION THROUGH RNA SPLICING.

These findings clearly show how studies on nuclear RNA complement gene cloning and sequencing approaches in revealing the structure and expression of active transcription units in eukaryotic cells. These studies clearly provide some definition of the steps and RNA species in the processing pathways to several immunoglobulin mRNAs in myeloma cells. However, it is unlikely that these immunoglobulin genes described here are regulated through RNA splicing mechanisms. First of all, these particular immunoglobulin genes are found in myeloma cells in which a substantial fraction (10-20%) of the total cell protein synthesized is the particular immunoglobulin product of these various cell lines. Quantitative studies on MOPC 21 κ mRNA processing indicate that all (>90%) of the κ mRNA sequences transcribed are conserved during nuclear RNA processing and ultimately transported to the cytoplasm as κ mRNA (3). Estimates of the transcriptional activity of this κ gene indicate that it approaches the most active estimates for rRNA genes in animal cells (3). It is likely that the other immunoglobulin genes considered here are also as actively transcribed and their transcripts as efficiently processed into cytoplasmic mRNA as the MOPC 21 κ mRNA. Thus, in retrospect these fully differentiated immunoglobulin-producing systems are likely to be poor candidates for studying the mechanisms controlling immunoglobulin gene expression.

It remains to be determined whether RNA splicing or other post-transcriptional RNA processing events are involved in regulating immunoglobulin gene expression during the developmental transitions in the immune response. However, antibody-producing cells undergo several transitions or switches in immunoglobulin gene expression during the development of the immune response which appear likely to be mediated through DNA rearrangements (see Hood, This Symposium) and may be regulated through differential RNA processing. One such transition is the shift from receptor IgM present on the surface of lymphocytes to IgM that is secreted from lymphocytes. Another is the switching of immunoglobulin heavy chain classes (i.e., C_H-regions) which retain the same V_H-region. We are now applying the RNA mapping techniques described here to these important and exciting questions in immunology and eukaryotic molecular biology.

ACKNOWLEDGMENTS

These studies were supported on NIH Grant, AI13410 and Program Project, CA12800. MGH was supported on a Lievre Senior Fellowship of the California Division of the American Cancer Society.

REFERENCES

1. Darnell, S.E., Progress in Nuc. Acid Res. Mol. Biol. 22: 327 (1978).

2. Abelson, J., Ann. Rev. Biochem. 48: 1035 (1979).

3. Gilmore-Hebert, M., and Wall, R., J. Mol. Biol. 135 (in press, 1979).

4. Wall, R., Gilmore-Hebert, M., Higuchi, R., Komaromy, M., Paddock, G., and Salser, W., Nuc. Acids Res. 5: 3113 (1978).

5. Rogers, J., Clarke, P. and Salser, W., Nuc. Acids Res. 6: 3305 (1979)

6. Sakano, H., Rogers, J.H., Huppi, K., Brack, C., Traunecker, A., Maki, R., Wall, R., and Tonegawa, S., Nature 277: 627 (1979).

7. Seidman, J.G., Max, E.E. and Leder, P., Nature 280: 370 (1979).

8. Gilmore-Hebert, M., Hercules, K., Komaromy, M., and Wall, R., Proc. Natl. Sci. USA 75: 6044 (1978).

9. Bailey, J.M. and Davidson, N. Anal. Biochem. 70: 75 (1976).

10. Alwine, J.C., Kemp, D. J., Parker, B., Reiser, J., Renart, J., Stark, G. and and Wahl, G., in Methods in Enzymology (in press, 1979).

11. Komaromy, M., Choi, E., Clarke, P., Kuehl, M., and Wall, R., in Eukaryotic Gene Regulation, Academic Press, New York (in press, 1979).

12. Seif, I., Khoury, G., and Dhar, R. Nuc. Acid Res. 6: 3387 (1979).

13. Ro-Choi, T.S., and Henning, D., J. Biol. Chem. 252: 3814 (1977).

14. Zieve, G. and Penman, S. Cell 8: 19 (1976).

15. Rogers, J. and Wall, R. Proc. Natl. Acad. Sci. USA (Manuscript submitted, 1979).

16. Franklin, E.D. and Frangione, B. in <u>Contemporary Topics in Molecular Immunology</u> <u>4</u>: 89 (1976).

17. Honjo, T., Obata, M., Yamawaki-Kataoka, Y., Kataoka, T., Kawakami, T., Takahashi, N., and Mano, Y. Cell <u>18</u>: 559 (1979).

18. Secher, D.S., Milstein, C., and Adetugbo, R. Immunol. Rev. <u>35</u>: 51 (1977).

19. Sakano, H., Huppi, K., Heinrich, G. and Tonegawa, S. Nature <u>280</u>: 288 (1979).

(See DISCUSSION on following page)

DISCUSSION OF DR. WALL'S PRESENTATION

Dr. Tom Wegmann: I do have a prediction for you Randy, and that is I wonder if you could make specific antibodies, preferably hybridoma antibodies, to the sequence of RNA that you are postulating to have the correct sequence for the splicing enzyme attached to it, and inhibit splicing with those antibodies.

Dr. Wall: There is one observation that I just learned about which Joan Steitz and her colleagues have made. If they precipitate a sonicate of nuclear RNA and chromatin with serum from lupus patients, she obtains a 10S RNP complex which has small RNAs in it. One of these appears to be U-1. Accordingly, the antibody you are talking about may be available.

Dr. Stavitsky: Would you explain the molecules made and the various kinds of heavy chains on the basis of splicing defects?

Dr. Wall: Let's say that some of them will be splicing defects. However, one alpha heavy chain variant which we have looked at which lacks the C_H3 domain in the protein does have the C_H3 sequence in the mRNA, and is apparently a nonsense mutation.

Dr. Knight: The primary transcript of the MPC 11 heavy chain presumably does not have information for μ but does it have information for the other gamma chains or for alpha chain?

Dr. Wall: No myeloma cell that we have looked at yet has a primary transcript that contains another heavy chain C region.

Dr. Hood: If your hypothesis is true, would it predict that you should have a certain order in removal of these intervening sequences? For example, in your light chains it always seems that the VC intervening sequence is saved for last.

Dr. Wall: Next to last in MPC 11.

Dr. Hood: But my question is, with your model it would seem to me that you could do any splicing at a time, or does that have to do with the secondary nature?

Dr. Wall: Well, there are two answers to that. One of them is that secondary structure of the RNA determines the timing. The other one is to propose, as I did, that the extent of the binding to the U-1 RNA determines the point in the processing pathway where the splice will occur. I should also point out one other element to the model. The model tells you how U-1 RNA might distinguish or recognize splicing sites and bring them into close proximity. What it does not do is tell you how U-1 RNA ignores splicing sites. Clearly, in the intervening sequence between the MOPC 21 V region and C region there are J regions, and those J regions have a very fine upstream splice site and a very fine downstream splice site which should be recognized by U-1 RNA. However, our kinetic data and our study of the nuclear precursors clearly indicate that thes splice sites are not used. By not used, I mean they are not used at least 90% of the time.

So the proposal basically advances a way in which U-1 RNA can distinguish and recognize splicing sites but unfortunately, I haven't come up with an idea which tells me how U-1 RNA may ignore splicing sites.

Unidentified Person: Have you looked at other cap structures because I think there are similarities between different cap sequences, and you might expect that the U-1 sequence would be found in other ends. Secondly, have you ever looked for the presence of more than one variable region in nuclear RNA?

Dr. Wall: I have not looked at cap structures in messenger RNA. The cap in U-1 RNA is different from the cap in messenger RNA though. It has an extra methyl in the G and it also has 2 secondary methylated bases.

To your second question; we have looked in MOPC 21 using the MOPC 321 and MPC 11 V-Regions which do not cross hybrizize with each other, and we have not found them. On the basis of our mapping data, we would have predicted that if there were another MOPC 21-like V region in the MOPC 21 K transcription unit, we would have obtained a complex curve for the inactivation of V region transcription. Obviously we do not see it, suggesting that there is only a single

MOPC 21 V region in this transcription unit.

Unidentified Person: How about the other variable regions that are far away. They perhaps could be transcribed in to some sort of precursor which is never processed.

Dr. Wall: We have never seen those in myeloma cells, just as we have never seen other heavy chain classes. My guess is that once you get to a myeloma cell, its genes have been activated such that they are first of all transcribed at about the same rate as ribosomal genes, (i.e. with several hundered polymerases transcribing a single gene). So they are very active genes, efficiently producting very abundant messengers.

Dr. Hood: Is there any correlation of the concentration of these small RNAs with the amount of processing that is going on in the cell? That is, do plasma cells which synthesize very large quantities of protein have more of these small RNAs?

Dr. Wall: That is a good point. Estimates suggest that every gene in a cell could have at least 10^6 U-1 RNA molecules. These are repeated genes apparently present several hundred times in the gene. The U-1 mRNAs are extremely stable (i.e., they are apparently as stable as ribosomal RNA). Because their small RNA's are present at an extremely high level, it appears that their level is not correlated with the level of gene expression. So, if there is regulation at the level of RNA splicing, it probably won't be by the limitation of this effector molecule.

Dr. Stavitsky: Have you looked for any messenger molecules or immunoglobulins in any T cell lines?

Dr. Wall: I do not have any T cell lines but I would be glad to do that if somebody would give me some.

Dr. Knight: Is it conceivable that you will be able to isolate these small molecules before the transcript is processed and then do the mixing experiments and see if you can actually show processing or degradation of the molecule.

Dr. Wall: We have been trying to think of clever ways to prove the involve-

ment of small RNAs in processing. We thought of trying to use _in vitro_ cross

linkers but that is not easy to do. I am hoping that the antibodies in lupus

serum will allow isolation of the splicing complexes in such a way that they

can be experimentally dissected.

THE SOURCES OF KAPPA LIGHT CHAIN DIVERSITY: A REVIEW

J. G. SEIDMAN, EDWARD E. MAX, BARBARA NORMAN, MARION NAU AND PHILIP LEDER
Laboratory of Molecular Genetics, National Institute of Child Health and
Human Development, National Institutes of Health, Bethesda, Maryland 20205 USA

SUMMARY

A model describing the arrangement of immunoglobulin light chain genes in
germline and antibody-producing cell DNA is presented. Mouse germline DNA
contains 300-1000 variable region genes each with a closely associated leader
segment, and five kappa J region segments closely associated with the one kappa
constant region gene. A site specific recombination event joins one germline
variable region gene with one of the germline J region genes to form an active
gene. Antibody diversity has two sources in germline DNA (many variable region
genes and five J region genes) and at least one source during somatic differentiation (the recombination event itself). Some of the data that led to the
formulation of this model is presented.

INTRODUCTION

A complete explanation of antibody gene expression must take into account
two critical observations--the striking variable/constant structure of antibody
molecules and the diversity of their sequences. How can the genetic information required to encode the millions of antibodies found in the serum of higher
animals be arranged without using a major portion of the mammalian genome? In
1965, Dreyer and Bennett first proposed that the variable and constant parts
of the light chain are encoded by two separate genes--a variable region gene
and a constant region gene.[1] Since then a variety of molecular biological
techniques have been applied to investigate the precise organization of these
genes. Most recently the advent of recombinant DNA technologies has permitted
a fairly comprehensive picture of the immunoglobulin light chain genes to
emerge.[2-10] Certainly a major source of antibody diversity must be the random
assortment of thousands of light chains with thousands of heavy chains. However, in what follows we will only describe how mouse DNA can encode the thousands of light chains without using a major portion of their genomes for encoding immunoglobulins.

A model of the arrangement of kappa light chain genes can be drawn from
the data presented elsewhere[2-10] and is summarized in Figure 1. Kappa light

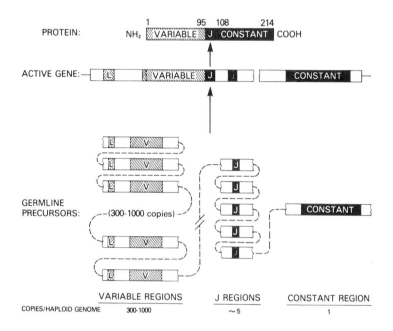

THE ARRANGEMENT AND ASSEMBLY OF
AN IMMUNOGLOBULIN LIGHT CHAIN GENE

Fig. 1. A model for the arrangement of immunoglobulin light chain genes. The light chain (214 amino acids long) is encoded by an active gene found only in immunoglobulin producing cells. This active gene is formed by a site-specific recombination event that joins a variable region to a J region segment. The two sources of germline diversity are the 300-1000 variable region genes and the five kappa J region segments.

chains are encoded in four segments in germline DNA.[2-10] The amino terminal half of the light chain is encoded by a leader and closely linked variable region (amino acids 1-95). The five J regions (encoding amino acids 96-108) and the single constant region (amino acids 109-214) are encoded in another region of the chromosome separated by a few thousand base pairs of DNA. An unusual feature of the light chain genes is that they require a site-specific somatic recombination event that joins a germline J region gene, to a germline variable region gene, to create an active gene. These recombinations occur during the differentiation of an antibody-producing cell. This model for the

arrangement of kappa light chains suggests that there are at best three sources
of antibody diversity--two germline and one somatic. Diversity encoded into
the germline DNA of Balb/c mouse resides first in 300-1000 kappa variable
region genes and second in five kappa J region segments (four of which are
active). These gene segments can generate between 1200 and 4000 variable region genes by the random assortment of any variable region gene with any J
region gene. The somatic recombination event that joins variable region genes
to J region genes generates diversity by allowing this random assortment of
germline gene segments. Furthermore, this somatic recombination event provides
a source of somatic diversification at a particular position in the kappa light
chain--amino acid 96. Many of the experiments that led to the formulation of
this model have been carried out independently in several laboratories (in particular the laboratory of Dr. S. Tonegawa), however we will refer here primarily to experiments we have done. The purpose of this article is to present
this model for the arrangement of kappa light chain genes and review the data
that led to its formulation.

The Procedure for Characterizing a Gene Segment

   The arrangement of the kappa light chain genes was determined by a variety
of molecular biological techniques that involve the identification and subsequent molecular cloning of restriction fragments, obtained from germline and
myeloma DNA, that encode segments of the kappa light chain. A prerequisite
for the application of these technologies is the availability of molecular
probes for the genetic segments of interest. In the case of the kappa light
chain these probes were obtained by the molecular cloning of light chain mRNA
sequences into bacterial plasmids. The methods for converting mRNA sequences
into cDNA, double-stranded cDNA and its subsequent insertion into bacterial
plasmids have been amply described (see ref. 11 for a review). The cloned mRNA
sequences were then cleaved by restriction enzymes into segments corresponding
to portions of the kappa light chain. These variable and constant region segments were used as hybridization probes to identify fragments of germline DNA
that encoded these portions of the light chain. Once identified, these fragments of germline DNA were cloned in bacteriophage lambda and characterized by
electron microscopic analysis and nucleotide sequence analyses.

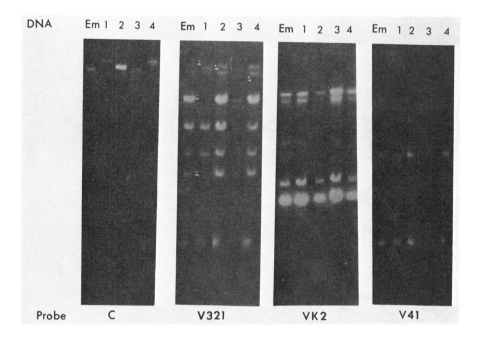

DNA

Em 1 2 3 4   Em 1 2 3 4   Em 1 2 3 4   Em 1 2 3 4

Probe   C   V321   VK2   V41

Fig. 2. One-dimensional analysis of EcoRl fragments of embryo and plasmacytoma DNAs annealed to constant and variable region probes. Each panel represents the EcoRl fragments of embryo[1], and five plasmacytoma DNAs fractionated by agarose gel electrophoresis and then blotted onto nitrocellulose filters.[12] Each filter was hybridized to ^{32}P-labelled probes corresponding to constant region or MOPC-321 (321), MOPC-149 (K2) or MOPC-41 (41) variable region sequences. The significance of band intensity is not yet clear, and may be a complex function of many factors. The most important factor is likely to be the degree of homology to the probe sequence.

The 300-1000 Mouse Kappa Leader/Variable Region Fragments Can Be Grouped Into 50-100 Sub-Families

Mouse kappa constant and variable region probes were used to identify a variety of restriction fragments of mouse DNA by the 'Southern' blotting techniques (Figure 2). In this procedure mouse DNA cleaved by restriction enzymes was fractionated on agarose gels, denatured, transferred to nitrocellulose filters, hybridized to ^{32}P labeled DNA probes; the filters were then subjected to autoradiography. While kappa constant region probes are able to find only a single EcoRl restriction fragment of mouse DNA, variable region probes are able to identify many restriction fragments encoding kappa variable regions. At least two explanations could be presented for the finding that a variable region gene probe identifies 6-10 EcoRl restriction fragments of mouse DNA. Either parts of the variable region could be found on many fragments, or each fragment could contain a slightly different kappa variable region gene. In order to distinguish between these two explanations, fragments that hybridized to variable region probes were cloned in bacteriophage lambda vectors and the nucleotide sequences of the variable regions encloded by the cloned fragments were determined.[5,7]

Variable region probes directed against variable regions from different families identify different sets of restriction fragments (Fig. 2 and ref. 7). (A variable region family was first defined by McKean et al. as the set of kappa light chains that differ by no more than three amino acids in the first 23 residues of the variable region.[28]) For example, the results presented in Figure 2 suggest that three different variable region gene probes from three different families identify three different sets of variable region gene containing fragments. Each variable region gene probe identifies 6-10 fragments. Several of the variable region containing restriction fragments from different families have been cloned and characterized. All of the fragments that have MOPC-321-like variable regions have been cloned[6], and two fragments each from the K2[7] and MOPC-41 (Fig. 3) families have been cloned and their variable region sequences determined. The nucleotide sequences of two MOPC-41-like variable regions, found on two of these restriction fragments, is presented in Figure 3. A complete variable region gene is found on each fragment, and each variable region has its own leader segment separated by a small intervening sequence. Although the variable regions of a family are similar, they are not identical.[5,7] For example, the MOPC-41 and the MOPC-173B variable regions differ by 14 of 95 amino acids. These and other variable region sequence comparisons suggest that under these conditions of hybridization a variable region probe will identify variable regions that differ from one another by 10-20%. Since there are ten mouse DNA restriction fragments that hybridize to the MOPC-41 variable region probe (Fig. 2 and ref. 5,7), this suggests that there are ten MOPC-41-like variable regions in the mouse genome.

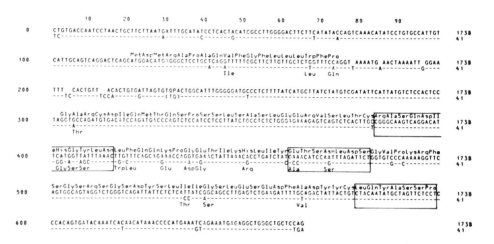

COMPARISON OF V REGION GENES: MOPC173B VS MOPC41

Fig. 3. A comparison of the MOPC-41 and MOPC-173B variable regions. The MOPC-41 variable region sequence has been presented elsewhere.[2] The MOPC-173B variable region sequence has been determined by Dr. E.E. Max (manuscript in preparation). The nucleotide sequences that are found in both variable region genes are indicated by '— —'. Differences between the variable region genes and the protein sequences that they would encode are also indicated. The nucleotide sequences were determined as described previously[2,8] using the procedure first described by Maxam and Gilbert.[23]

Each variable region is thought to have an associated leader segment for two reasons. First, the five variable region genes whose nucleotide sequences have been determined all have associated leader segments (ref. 7 and Fig. 3). Second, a leader segment is found on every fragment that has a variable region. That is, probes have been fashioned from cloned segments of genomic DNA that encode the MOPC-41 and MOPC-321 leader segments but not the MOPC-41 or MOPC-321 variable region genes, these ' leader' probes identify all of the fragments identified by MOPC-41 or MOPC-321 variable region gene probes. The Southern blots with leader probes are identical to Southern blots with variable region probes (J.G.S. and P.L., unpublished results). Thus there is a leader segment associated with each kappa variable region gene.

How many mouse kappa leader/variable genes are encoded in germline DNA? Analysis of the kappa light chains secreted by mouse plasmacytomas suggests that the mouse must be able to make light chains from 50-100 families (Dr. M. Potter, personal communication). If each family is encoded by 6-10 members then there are 300-1000 variable region genes in the mouse genome. Furthermore, preliminary data suggests that some of the variable region genes of the MOPC-41 family are separated from one another by about 10 kb (Dr. Henry Miller and P.L., unpublished results); if every variable region gene is separated from its nearest neighbor by 10 kb then the kappa light chain variable region genes will occupy between 3 and 10 million base pairs of DNA or about 0.2% of the total mouse genome.

The Germline Kappa Constant Region Gene is Closely Linked to the Kappa J Region Segments

A variety of genetic experiments as well as kinetic hybridization analyses all suggested that there is only a single kappa constant region gene in the haploid mouse genome. As with the variable region gene sequences recombinant DNA technologies have permitted the isolation and characterization of kappa constant region genes.[2,14]

The nucleotide sequence of the mouse kappa constant region gene and about 5000 bp of DNA found to the 5' side of the gene have been determined (E.E.M. and P.L., unpublished results). The nucleotide sequence of the gene suggests that it encodes amino acids 109-214 on a single uninterrupted stretch of DNA. However, as can be deduced from the nucleotide sequences shown in Figures 3 and 4, the variable region and constant region genes cannot encode the entire light chain. Another segment, the J segment, is required to encode amino acids 96-108 of the light chain. The J segment was first identified in the lamba light chain gene system.[3,14] The kappa light chain system also uses J segments. Five germline kappa J regions have now been identified to the 5' side of the kappa constant region (Fig. 4). These five kappa J region segments are encoded in a stretch of DNA separated from one another by about 300 base pairs between 2.5 and 3.9 kb from the kappa constant region gene.

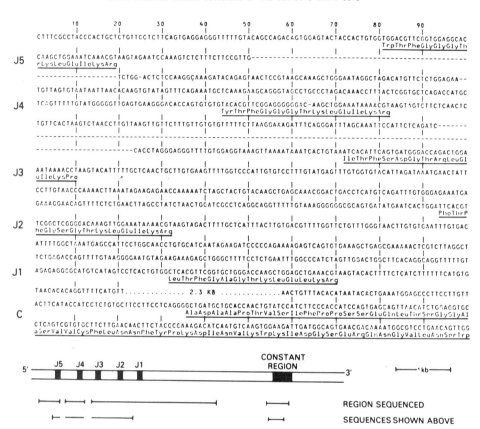

J AND CONSTANT REGION SEQUENCES OF THE GERMLINE KAPPA GENE

Fig. 4. Sequence of the region containing kappa J- and C-region genes. Coding regions of the genes are indicated by the amino acids below the nucleotide sequence. The dashes indicate undetermined nucleotides. The lengths of these unsequenced regions were estimated by restriction analysis. The diagram below indicates the extent of the sequence determined and shows which portion of the known sequence is given above. The J regions were identified by their homology with the J of the expressed MOPC-41 gene (J5). They are numbered from the J closest to the C region. These data have been published previously.[3]

There are probably only five kappa J region segments in the entire mouse genome. The evidence for this statement is circumstantial and comes from four sources. First, there are no other kappa J region segments on the cloned EcoRl fragment that contains the kappa constant region gene since the entire nucleotide sequence of the 4.6 kb to the 5' side of the constant region gene is now available (E.E.M. and P.L., unpublished results) and this sequence encodes only five kappa J region segments. Second, when the J3/J4 sequences are used as a

hybridization probe they anneal only to themselves--these probes do not identify other kappa J region segments on other restriction fragments (ref. 15 and
our unpublished results). Third, four kappa J region segments can account for
all of the J regions found in Balb/c myeloma proteins whose amino acid sequences are available.[3,4] Fourth, Perry et al.[29] have shown that the sizes of RNA
precursors for a large number of NZB mouse light chains are consistent with the
notion that there are only four active J regions. In fact although five J
segments are found in the DNA only four of these are actually expressed in
myeloma proteins. J3 (Fig. 4) is probably defective for at least two reasons
and cannot be expressed as part of a light chain.[3] A comparison of these J
region segments to one another[3,4] and the lambda J region[2] has already lead
to the identification of conserved DNA sequences that may play a role in V/J
recombination. Further comparisons of J regions from other mouse strains, and
other mammals will show how these segments have altered during evolution.
Eventually, we will know how much variation of the J regions has occurred and
this should indicate the contribution of the J region to light chain diversity.

The Somatic Joining of a Variable and a J Region Gene Segment Creates an Active Gene

Since Dreyer and Bennett[1] first proposed the two gene-one polypeptide model
geneticists have wondered how an immunoglobulin light chain gene is activated
in an antibody producing cell. Apparently, the expression of a particular
variable region gene requires recombination with a J region segment to form an
active gene with the structure shown in Figure 1. Such "V-J recombination"
was first characterized in the lamdba light chain gene system[13,14] and more
recently has been shown to be required for the formation of kappa light
chains[2-10] and immunoglobulin heavy chains.[16-19] Presumably, the signals for
somatic DNA recombination are encoded to the 5' side of the germline J region
gene and the 3' side of the germline variable region gene. The finding that
there are conserved sequences to the 5' side of each kappa J segment and the
3' side of each kappa variable region gene supported this belief.[3,4,8] A role
for these conserved sequences in the joining of variable region to J region
gene segments has been proposed.[3,4] However, in order to understand the exact
nature of one of the recombination events, the DNA fragment encoding the two
germline precursors and the active gene must be characterized.

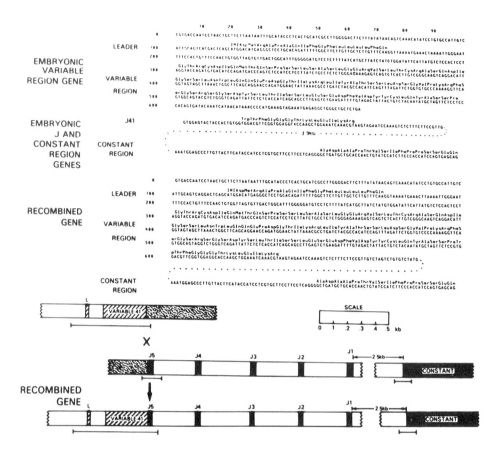

Fig. 5. Nucleotide sequences of the germ line and expressed genes encoding the MOPC-41 light chain. a, Germ line leader and variable region gene; b, germ line J region and constant region genes; c, the leader/variable gene expressed in the MOPC-41 cell has been formed by fusing the germ line variable and J region genes. The location of each sequence with respect to the structural gene sequences is indicated on the figure below each sequence. The locations of four additional J-region genes, J4, J3, J2 and J1 in the intervening sequence were determined by Max et al.[3] using nucleotide sequence analysis.

The recombination event that led to the formation of the gene expressed in the MOPC-41 myeloma can now be studied in some detail since all three fragments--two from germline DNA and one from myeloma DNA--have been cloned.[8] Figure 5 shows the nucleotide sequences of the MOPC-41 variable region in its germline and active gene contexts. The MOPC-41 variable region gene has fused directly to a germline J region, J5, to create an active gene.

The creation of an active gene by V-J recombination suggests a further mechanism for increasing antibody diversity. That is, variable region and and J region segments can probably assort randomly.[20] If this is true, then the use of four active kappa J regions (see below) will lead to at least a 4-fold increase in the number of expressed light chains over the germline variable region gene repertoire.

Light Chain Diversity is Further Increased by V-J Recombination

One might have thought that the somatic recombination event that creates an active gene would be a precise event. Since V-J recombination occurs in the middle of a structural gene segment and since the total number of nucleotides in the active gene must always be the same, then perhaps the recombination event would always occur in the same position of the variable and J region gene segments. However, both direct and indirect evidence suggest that the recombination event is imprecise, the location of the recombination site varies relative to the variable region and J region genes, from one recombination event to the next.

The indirect evidence for this shift in the location of the recombination site comes from an analysis of the J regions that have been sequenced in Balb/c myeloma proteins (Fig. 6 and refs. 3,4). The germline J region segments can encode all of the expressed J regions except at amino acid position 96 (Fig. 6). In fact, position 96 of the light chain is one of the most variable positions in the light chain.[21] Since position 96 is precisely at the V-J boundary and since all of this diversity cannot be germline encoded, the diversity found at this position is probably generated by the V-J recombination event.

A mechanism to explain this somatic generation of diversity based on the nucleotide sequences of the kappa variable and J region gene segments has been proposed (Fig. 7). If the site of V-J recombination can shift from one position to the next then this could generate amino acid diversity at position 96 (Fig. 7). In fact this model could account for all of the diversity that has been found at position 96 in Balb/c myeloma proteins.

Two examples of shifts in the site of V-J recombination have recently been identified (E.E.M., J.S., P.L., Drs. S.P. Kwan and M. Scharff, manuscript in preparation). There are now two examples of recombined genes, cloned from myeloma DNA, where a variable region has been fused to a J region, but nucleo-

AMINOACID TRANSLATION OF GERMLINE
BALB/c J-REGION GENES

A.	W T F G G G T K L E I K R*	J5
B.	Y - - - - - - - - - - - -*	J4
C.	I - - S D - - R - - - - P*	J3
D.	F - - - S - - - - - - - -*	J2
E.	L - - - A - - - - - L - -*	J1

AMINOACID SEQUENCES OF J-REGIONS FOUND
IN KAPPA CHAINS OF BALB/c MICE

	96 108	
#1.+	W T F G G G T K L E I K R*	J5
#2.	R - - - - - - - - - - - -	
#3.	Y - - - - - - - - - - - -*	J4
#4.	P - - - - - - - - - - - -	
#5.	F - - - S - - - - - - - -*	J2
#6.	W - - - S - - - - - - -	
#7.	I - - - S - - - - - - -	
#8.	L - - - A - - - - - L - -*	J1
#9.	I - - - A - - - - - L - -	

* = GERMLINE SEQUENCE

+ The myeloma proteins used as examples are identified and referenced as follows: #1 MOPC 41 (21), #2 MOPC 173 (21), #3 MOPC 21 (24), #4 MOPC 11 (25), #5 MOPC 149 (J.G. Seidman, unpublished results), #6 MOPC 321 (36), #7 X 24 (27), #8 MOPC 511 (E. Apella, personal communication), #9 TEPC 601 (27).

W = Trp, T = Thr, F = Phe, G = Gly, K = Lys, L = Leu, E = Glu,
I = Ile, R = Arg, Y = Tyr, P = Pro, S = Ser, D = Asp, A = Ala

Fig. 6. Germline J-region gene translations and observed J-region amino acid sequences.

tides have been lost during the recombination event. Apparently, the V-J recombination event is rather imprecise and this imprecision generates further antibody diversity.

Are There Other Mechanisms to Generate Antibody Diversity?

Antibody diversity is encoded by multiple germline variable region and J region genes and is further generated by the somatic V-J recombination event. All of the sources of kappa light chain diversity encoded in the germline DNA have probably been identified. However, further light chain diversity could be generated by specific somatic diversification mechanisms. That is, during the somatic differentiation of an antibody producing cell, the variable region

ALTERING THE FRAME OF SOMATIC RECOMBINATION ACCOUNTS FOR DIVERSITY AT AMINO ACID 96

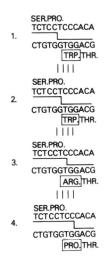

Fig. 7. Altering the site of somatic recombination accounts for diversity at amino acid 96. In each example the nucleotide sequence at the 3' end of the MOPC-41 variable region is shown aligned above the 5' side of J5. Four different recombination sites are used and the boxed amino acid is residue 96 of the light chain. Notice that if the genes recombine as indicated in example 3 or 4, different amino acids are encoded at residue 96.

gene might be mutated by a specific enzymatic mechanism to further increase the light chain repertoire. Direct evidence for somatic mutational events have been identified among the lambda light chains.[22] Weigert et al.[20] have suggested that such somatic mutation of the kappa light chain genes can also occur. However, results of nucleotide sequence analyses of germline and recombined kappa variable region genes (for example, Fig. 5 and ref. 2) suggest that such somatic mutation does not extensively alter the variable region gene. To date, evidence for somatic variation of light chain variable region genes has only come from an analysis of myeloma proteins and their active genes. Two questions remain: 1) Does the somatic variation identified among myeloma proteins reflect a physiologically important phenomena? and 2) Is there a special enzymatic mechanism for generating this somatic variability? Eventually, analyses of recombined genes isolated from spleen cells, and the germline precursors of these recombined genes will provide an answer to the first question and may provide clues to the second question.

REFERENCES

1. Dreyer, W.J. and Bennett, J.C. (1965) Proc. Natl. Acad. Sci. U.S.A. 54, 864
2. Seidman, J.G. and Leder, P. (1978) Nature 276, 790.
3. Max, E.E., Seidman, J.G. and Leder, P. (1979) Proc. Natl. Acad. Sci. U.S.A. 76, 3450.
4. Sakano, H., Huppi, K., Heinerich, G. and Tonegawa, S. (1979) Nature 280, 288.
5. Seidman, J.G., Leder, A., Edgell, M.H., Polsky, F., Tilghman, S.M., Tiemeier, D.C. and Leder, P. (1978) Proc. Natl. Acad. Sci. U.S.A. 75, 3881
6. Lenhard-Schuller, R., Hohn, B., Brade, C., Hirama, M., and Tonegawa, S. (1978) Proc. Natl. Acad. Sci. U.S.A. 75, 4709.
7. Seidman, J.G., Leder, A., Nau, M.M., Norman, B. and Leder, P. (1978) Science 202, 11.
8. Seidman, J.G., Max, E.E. and Leder, P. (1979) Nature 280, 370.
9. Bernard, O., Hozumi, N. and Tonegawa, S. (1978) Cell 15, 1133.
10. Nishioka, Y. and Leder, P. (1980) J. Biol. Chem. 255, 3691.
11. Genetic Engineering, CRC Press (1978) West Palm Beach, Fla., eds. A.M. Chakraborty.
12. Southern, E.M. (1975) J. Mol. Biol. 98, 503.
13. Bernard, D., Hozumi, N. and Tonegawa, S. (1978) Cell 15, 1.
14. Tonegawa, S., Maxam, A.M., Tizard., R., Bernard, O. and Gilbert, W. (1978) Proc. Natl. Acad. Sci. U.S.A. 75, 1485.
15. Perry, R.P., Kelley, D.E., Coleclough, C., Seidman, J.G., Leder, P., Tonegawa, S., Mathysseus, G., Weigert, M. (1980) Proc. Natl. Acad. Sci. U.S.A. 77, 1937.
16. Early, P., Huang, A., Davis, M., Calame, K. and Hood, L. (1980) Cell 19, 981.
17. Sakano, H., Rogers, J.H., Hüppi, I., Brack, C., Traunecker, A., Maki, R., Wall, R. and Tonegawa, S. (1979) Nature 277, 627.
18. Kataoka, T., Yamawaki-Kataoka, Y., Yamagishi, H. and Honjo, T. (1979) Proc. Natl. Acad. Sci. U.S.A. 76, 4240.
19. Kemp, D.J., Cory, S. and Adams, J.M. (1979) Proc. Natl. Acad. Sci. U.S.A. 76, 4627.
20. Weigert, M., Gatmaitan, I., Loh, E., Schilling, J. and Hood, L. (1978) Nature 276, 785.
21. Kabat, E.A., Wu, T.T. and Bilofsky, H. (1979) Sequences of Immunoglobulin Chains.
22. Weigert, M.G., Cesari, I.M., Yonkovich, S.J., and Cohn, M. (1973) Nature 288, 1045.

23. Maxam, A.M. and Gilbert, W. (1977) Proc. Natl. Acad. Sci. U.S.A. 74, 560.
24. Padlan, E.A., Davies, D.R., Pecht, I., Givol, D. and Wright, C. (1976) Cold Spring Harbor Symp. Quant. Biol. 41, 627.
25. Rao, D.N., Rudikoff, S., Krutzsch, H. and Potter, M. (1979) Proc. Natl. Acad. Sci. U.S.A. 76, 2890.
26. Breathnach, R., Benoist, C., O'Hare, K., Gannon, F. and Chambon, P. (1978) Proc. Natl. Acad. Sci. U.S.A. 75, 4853.
27. Catterall, J.F., O'Malley, B., Robertson, M.A., Staden, R., Tanaka, Y. and Brownlee, G.G. (1978) Nature 275, 510.
28. McKean, D.J., Bell, M. and Potter, M. (1978) Proc. Natl. Acad. Sci. U.S.A. 75, 3913.
29. Perry, R.P., Kelley, D.E. and Schibler, U. (1979) Proc. Natl. Acad. Sci. U.S.A. 76, 3678.

(See DISCUSSION on following page)

DISCUSSION OF DR. SEIDMAN'S PRESENTATION

Dr. Wall: How similar are the nucleotide sequences of the kappa genes of mouse and human?

Dr. Seidman: They are 70% homologous and the boundaries are in exactly the same position.

Dr. Hood: In a myeloma, wherein two kappa C regions had undergone rearrangements, what kind of rearrangements did the second one undergo? Dit it also join to V?

Dr. Seidman: Sometimes, but sometimes you would find a completely barren thing. For example, the V region may be brought very close to the C region. In some cases, you find that there is no V region there at all and in still others, you find very close to the normal recombination has occurred.

Unidentified Person: Is there any homology in the signal sequences between the DNA and proteins in the mouse or isn't there enough information on this? Assuming that there are signal sequences involved in the joining of V to J and other segments for other proteins like globin, are there any similarities in the signal sequences for recombination?

Dr. Seidman: Between the light chain and the heavy chain, the sequence that Dr. Hood pointed out seems to be conserved. It is also conserved between lambda and kappa.

Dr. M. Schiffer: We find that residue 96 in human kappa light chains is very important in that it seems to determine the association properties of these light chains. For example, if you have an arginine at position 96, the light chains don't form dimers, but if tryptophan or tyrosine is at position 96, they dimerize with a high dimerization constant.

Dr. Seidman: Do you have any evidence whether residue 96 of the kappa chain affects their ability to bind with heavy chains? That is, if you change position 96 in light chains are their certain classes of heavy chains that they do not combine with well?

Dr. M. Schiffer: We don't have any data on this but I would expect this to happen.

Dr. Knight: Counting the number of V region genes relies on the hybridization of your probe to the various restriction fragments and the assumption is that one V region probe will hybridize to other genes coding for molecules of that subgroup but not to genes coding for V regions of other subgroups. You showed us that kappa 41 and the kappa 149 do not cross-hybridize and I wonder how confident you are that this is the case throughout all of the subgroups? How different are the kappa 41 and kappa 149?

Dr. Seidman: Kappa 41 and 149 differ by about 50%. I think that in general the different subgroups differ about that much from one another.

Dr. Knight: So you feel confident then that you will not get cross hybridization?

Dr. Seidman: So far, every time we have used a different variable region probe from a different family we found that a different set of fragments hybridize and this has been done in 6 subgroups now.

Unidentified person: Is there an embryonic chromosome or non-rearranged chromosome in myeloma cells?

Dr. Seidman: In particular, in fact one has to look rather hard for the right cell to do it in. In turns out that the best thing to look at is cloned cell lines. For example that was S107 that had been cloned by Dr. Sharpin, and he actually provided us with the DNA which we were then able to analyze. It turns out that J3 and J4 have been lost from both chromosomes. All of the cloned cell lines that we have looked at so far appear to have completely lost the embryonic chromosome. Now, whether that is a function of the cloning event itself we really don't know. Certainly the myeloma lines that we have looked at do seem to have both chromosomes.

II
Immunology: Current Developments

TUMOR IMMUNOLOGY

JANE BERKELHAMMER[+] AND NICHOLAS M. PONZIO[++]
[+]Cancer Research Center, Department of Medicine, Division of Biological
Sciences, and Sinclair Comparative Medicine Research Farm, University of
Missouri, Columbia, Missouri 65201, USA; [++]Department of Microbiology-
Immunology, and Cancer Center, Northwestern University Medical School,
Chicago, Illinois 60611, USA

INTRODUCTION

What is tumor immunology? Woodruff[1] states that one definition of an immunological reaction is a response triggered by antigen and mediated by specific
antibody or sensitized lymphocytes. Yet, it is becoming increasingly clear
that present day studies of host tumor interactions cannot be restricted only
to antigen-specific immunologic responses. One reason for this is that tumor-
specific antigens have been extremely difficult to identify and virtually im-
possible to isolate from tumor cells. Secondly, "non-specific" responses
appear to abound in many tumor systems. Two major examples of seemingly non-
specific responses are tumor kill by natural killer (NK) cells and by activated
macrophages.

Finally, immunologic responses to tumors cannot be restricted to antigen-
specific responses, because tumors are heterogeneous populations of cells often
with diverse biologic and antigenic characteristics. The diversity of tumor
immunology was most effectively indicated by the range of topics presented and
discussed in this workshop. The workshop papers are summarized below together
with supplementary comments, where appropriate, highlighting related studies
of other investigators.

TUMOR HETEROGENEITY

Drs. F. Miller, B. Miller and G. Heppner presented some of their elegant
studies on mouse mammary tumor heterogeneity. These investigators and their
colleagues[2-4] have isolated and characterized several cell lines from a single
parent tumor that arose spontaneously in a BALB/cfC3H mouse. The isolated cell
lines differ with respect to growth rates, expression of tumor associated antigens and the ability to induce effective immunity. Further analysis of the
immunogenicity of two such sublines revealed an apparent state of concomitant
immunity. Injection of one cell line subcutaneously on one side of a BALB/

cfC3H recipient invokes an immune response that is effective in diminishing the growth of a second subline, which by itself is unable to induce an effective immune response. Although humoral immunity to these isolated sublines has not been assessed, cell-mediated immune responses as measured by the Winn assay, ^{51}Cr-release assay and microcytotoxicity assay indicate some cross-reactivity among five of the isolated sublines.

As Dr. Heppner has pointed out in a masterful review of tumor heterogeneity[5], the "heterogeneous nature of cancer is not a new concept," and the phenomenon was described in many reports of the 1950's and 60's, e.g.[6,7] Despite these early reports, many cancer investigators have fallen prey to the universal desire to reduce biological systems to simple, homogeneous models for study. In certain circumstances such an approach is desirable, even essential. However, in the study of host-tumor interactions, the heterogeneity of tumors is fundamental to their biologic behavior, thus, in any attempt to understand and control tumor growth, the heterogeneous nature of tumors must be carefully considered.

Cancer investigators are beginning to reexamine the problem of tumor heterogeneity, partly as a result of disappointing clinical results with therapy of most solid tumors. Several investigators have demonstrated that tumors are composed of subpopulations of cells with varying sensitivities to drugs,[8-10] and radiation.[11] Fidler et al.[12,13] have contributed greatly to the renewed interest in tumor heterogeneity with the isolation and characterization of metastatic variants from the murine B16 melanoma. Fidler and Kripke have demonstrated[14] that metastatic properties are contained in specific subpopulations within the parent tumor. Other investigators[15,16] have obtained similar results, and still others[17] have found immunologic heterogeneity between primary and metastatic tumors as well. These studies suggest the existence of antigenic diversity among cells in the primary tumor.

Cytotoxic effectors. The phenomenon of tumor heterogeneity can have a profound effect upon the generation of cytotoxic effector cells. In this workshop, for example, Drs. L. Simon, J. H. Finke and M. R. Proffitt discussed the characteristic properties of various in vitro cell lines derived from lymphomas that arose in Moloney murine leukemia virus (MuLV-M) infected C3H/He mice. The data presented indicate that lymphomas induced in the same strain of mice by the same agent, may represent transformed cells of different lymphocyte subpopulations that have different phenotypic expressions. Thus, the three cell lines investigated differ with respect to morphology, virus production and expression of antigenic determinants. Functionally, these lines also differ in their ability to stimulate production of syngeneic and allogeneic cytotoxic effector cells and in their tumorigenic potential. Of particular interest to the latter observation was the fact that two of the three lines studies expressed immuno-suppressive properties. Further investigations designed to identify additional functional characteristics (e.g., helper function, cytotoxic function) or MuLV-M transformed cell lines may provide homogenous populatons of lymphocyte sub-sets for a more detailed analysis of lymphocyte effector functions.

In this regard, Drs. K. L. Fitzgerald and N. M. Ponzio reported on a trans-plantable lymphoma of SJL/J mice that possesses effector cytotoxic function. Originally described as a reticulum cell sarcoma (RCS), this tumor has more re-cently been shown to be of B cell lineage. Among some very interesting aspects of the host-tumor relationship exhibited by this lymphoma, of particular rele-vance here is the observation that when used as effector cells in a ^{51}Cr-release assay, RCS tumor cells will lyse only those targets that are also susceptible to lysis by natural killer (NK) cells. RCS cells compare favorably with "conven-tional" NK cells regarding target cell range, lack of involvement of macrophages and lack of Thy-1 marker. While it is possible that RCS cells may represent a tumor of NK cells, the tumor cells display several characteristics which "con-ventional" NK cells lack, most notably a much greater resistance to gamma ir-radiation and the expression of I-region coded (Ia) determinants. Therefore, RCS cells may reflect an entirely new class of lymphoid cells with cytotoxic (NK) effector function.

Natural cell-mediated cytotoxicity has been found in several species, in-cluding rats, mice, and man. The demonstration of natural cell-mediated cyto-toxicity in tumor-bearing hosts may explain much of the "non-specific" reac-tivity formerly attributed to variabilities in in vitro assay systems.[18]

The mouse NK cell is generally thought to be a non-T cell, particularly be-cause of the high levels of reactivity in nude mice.[18] NK reactivity is usually unaffected by pretreatment of lymphoid cells from conventional or nude mice with anti-theta serum plus complement.[19-22] Similarly, pretreatment of NK cells with anti-kappa serum plus complement caused no inhibition of reactivity[23] and pas-sage of reactive cells over an anti-immunoglobulin column increased relative reactivity.[21] Human NK cells have also been studied extensively, but reports of cell surface markers are conflicting.[18]

NK cell activity may provide an alternate mechanism to the classical concept of immunologic surveillance as mediated by specifically sensitized T lymphocytes.

In another workshop paper, Drs. T. L. Bowlin and M. R. Proffit described experiments in which methylcholanthrene (MCA) transformed C3H/He fibroblasts were studied for their ability to stimulate cytotoxic effector cells. The MCA-induced cells readily produce tumors when as few as 10^3 are injected into immunocompetent adult syngeneic mice. However, using either syngeneic or allogeneic responder cells, MCA-transformed tumor cells fail to generate detectable cytolytic effector cells _in vitro_. These tumor cells are, however, susceptible to lysis when used as targets for allogeneic C3H-induced cytotoxic cells. It is suggested that the MCA-transformed clone of C3H fibroblasts may express some of the necessary antigenic determinants required for cell lysis, but may lack other determinants that are required to stimulate the lymphoid cells necessary to generate cytotoxic effectors. Preliminary evidence also suggests the possibility that this tumor cell may stimulate suppressor cells that influence (in a negative way) the production of cytotoxic cells. In either instance, the lack of production of an efficient immunity may account, in part, for their tumorigenic potential in immunocompetent recipients.

Nelson et al.[24] also reported a failure of methylcholanthrene-induced sarcomas to elicit the production of rapidly cytolytic T cells in syngeneic mice. These cells did elicit the production of more slowly cytolytic cells and were susceptible to kill by T cells from allo-immune mice and from virus-immune H-2 compatible mice. These investigators suggested that failure to elicit cytolytic T cell production might result from lack of an appropriate association between tumor-specific antigens and elements of the major histocompatibility complex. Oliver[25] in a review of the HLA system and its possible association with immunological defense against cancer, has suggested that an understanding of the mechanisms of action of HLA associated resistance factors may enable a more rational approach to immunotherapy in man.

Antigen-Induced Tolerance. Dr. N. Bhoopalam reported on factors that influence the growth of two plasmacytomas (MOPC 104E and J558) _in vivo_ and _in vitro_. Both tumors produce immunoglobulins that bind to α,1-3 dextran. The MOPC 104E cells possess receptors for dextran, however the J558 cells do not. Injection of dextran 7 days prior to or 3 days after tumor injection results in an inhibition of tumor growth. _In vitro_ proliferation (as judged by ^{3}H-thymidine incorporation) is suppressed in both tumors when anti-dextran anti-bodies are added to the culture medium, however, only the MOPC 104E tumor cells are inhibited by in vitro addition of dextran. The presence of spleen cells at a ratio of 20:1 (spleen:tumor) also suppress thymidine uptake by both tumors, however, normal spleen cells, as well as spleen cells from tumor bearing mice or from mice injected with dextran show this suppressor capacity. It is suggested that the suppression of tumor growth by specific antigen in this instance is comparable to antigen-induced tolerance and may provide a model system to study B cell tolerance.

UNIQUE ANIMAL MODELS

Ultraviolet light carcinogenesis. Blum[26] and Kripke[27] have demonstrated that skin tumors can be readily induced in mice with chronic UV irradiation. These tumors are extremely interesting immunologically because they are highly antigenic and are often rejected after transplantation to normal syngeneic recipients.[27,28] However, these tumors do grow progressively when transplanted to immunosuppressed or UV-irradiated mice.[29] The rejection of these tumors in immunocompetent hosts has been shown to depend on theta-bearing lymphocytes.[30,31]

Further, Kripke and Daynes and their associates[32-34] have shown that failure of UV-treated hosts to reject syngeneic UV-induced tumors results from the presence of suppressor T cells.

Daynes and co-workers have characterized the cell types involved in the generation of cytotoxic lymphocytes. For example, they demonstrated[35] the presence in UV-tumor tissue of significant numbers of host-derived macrophages which appear to be important to the in vitro generation of cytotoxicity. The cytotoxic effector cells generated appear to be active against all cell UV-induced tumors, as well as against tumors induced with benz[a]pyrene and methylcholanthrene.[36]

These investigators have demonstrated the presence on UV-induced tumors of both unique and common determinants which can function as rejection antigens in in vivo tumor transplantation assays.[37] UV-generated suppressor cells are apparently capable of specifically inhibiting effector responses directed against common tumor antigens.[37] Daynes et al.[38] have also shown that low dose gamma radiation, which selectively depresses suppressor T-cell function in vivo, appears to enhance tumor immunity by removing regulatory influences.

Thus, one explanation for the failure of "immune surveillance" in vivo might
be that responses to common tumor antigens are stronger and more readily recognized than those directed against tumor-specific antigens. Therefore, suppressor T cells directed against the common antigens might override the immune
response directed against tumor-specific antigens, and thus suppression of tumor
immunity would prevail in vivo. These elegant studies clearly demonstrate the
importance of immunoregulatory mechanisms in host-UV tumor interactions.

Sinclair swine melanoma. Sell has stated[39] that, although animal experiments
have often provided important insights into the mechanisms of human disease and
the rationale for therapy, extensive immunotherapy studies in animal systems
have not been reproducible in man. He further maintains that new animal models
which more closely reflect the properties of human cancer are needed to determine the true potential of cancer therapy in man.

One such new model, discussed by Drs. J. Berkelhammer and R. R. Hook in this
workshop, is spontaneous malignant melanoma that occurs in Sinclair miniature
swine. These tumors develop at birth or in young animals[40] and histopathologi-
cally resemble human superficial spreading melanoma.[41] There is a wide spectrum
of benign melanocytic lesions which are capable of malignant transformation, and
metastatic disease is not uncommon.[40] In addition, Hook et al.[40] have demonstrated that the incidence of swine melanoma can be increased by selective
breeding. This hereditary aspect of the disease may be comparable to familial
melanoma in man.[42]

Unlike human melanoma, however, swine melanoma exhibits a high rate (85%) of
spontaneous regression.[43] Preliminary studies by Berkelhammer, Hook and co-
workers[44-47] suggest that immunologic factors participate in this spontaneous
regression. The evidence for this comes from observations that (1) swine melanoma cells grow progressively when transplanted to immunologically deficient
environments such as cell culture[44] and the hamster cheek pouch;[45] (2) melanoma
swine which have been immunologically suppressed by thymectomy and antilympho-
cyte serum treatment exhibit increased tumor growth when compared to untreated
siblings;[46] and (3) a tumor-specific cell mediated immune response can be demonstrated in vitro using lymphocytes from tumor-bearing swine.[47]

Several factors suggest that tumor heterogeneity may also play a significant
role in the pathogenesis and regression of Sinclair swine melanoma. These
lesions exhibit a variety of morphologies both in vivo[41] and in vitro.[44] In
addition, several swine melanoma lesions can be seen progressing and regressing

simultaneously on a single swine. Thus the Sinclair swine melanoma system provides a unique opportunity to study host-tumor interactions in a clinically relevant model of neoplasia.

ACKNOWLEDGEMENTS

The authors thank Mrs. Dorothy Tumulty for typing the manuscript. Dr. Berkelhammer's research was supported by Grants CA-08023, CA-19409, and CA-25718 from the National Cancer Institute, DHEW.

REFERENCES

1. Woodruff, M. (1979) Br. J. Surgery, 66, 297.
2. Dexter, D. L., Kowalski, H. M., Blazar, B. A., Fligiel, Z., Vogel, R., and Heppner, G. H. (1978) Cancer Res. 38, 3174.
3. Heppner, G. H., Dexter, D. L., DeNucci, T., Miller, F. R., and Calabresi, P. (1978) Cancer Res. 38, 3758.
4. Miller, F. R., and Heppner, G. H. J. Natl. Cancer Inst. (In Press).
5. Heppner, G. H. in Commentaries on Research in Breast Disease. Bulbrook, R. D. and Taylor, D. Jane, ed., Alan R. Liss, New York (In Press).
6. Levan, A., and Hauschka, T. S. (1953) J. Natl. Cancer Inst. 14, 1.
7. Klein, G., and Klein, E. (1956) Ann. N.Y. Acad. Sci. 63, 640.
8. Barranco, S. C., Drewinko, B., Ho, D., Humphrey, R. M., and Romsdahl, M. M. (1972) Cancer Res. 32, 2733.
9. Barranco, S. C., Drewinko, B., and Humphrey, R. M. (1973) Mutation Res. 19, 277.
10. Hakannson, L., and Tropé, C. (1974) Acta Path. Microbiol. Scand. Section A, 82, 35.
11. Leith, J. T., Zeman, E. M., Heppner, G. H., and Glicksman, A. S. (1979) Proc. Am. Assoc. Cancer Res. 20, 28. (Abs.)
12. Fidler, I. J. (1973) Nature New Biol. 242, 148.
13. Fidler, I. J. (1975) Cancer Res. 35, 218.
14. Fidler, I. J., and Kripke, M. L. (1977) Science, 197, 893.
15. Tao, T. W., and Burger, M. M. (1977) Nature, 270, 437.
16. Suzuki, N., Withers, H. R., and Koehler, M. W. (1977) Cancer Res. 38, 3349.
17. Pimm, M. V., and Baldwin, R. W. (1977) Int. J. Cancer, 20, 37.
18. Herberman, R. B., and Holden, H. T. (1978) Adv. Cancer Res. 27, 305.
19. Herberman, R. B., Nunn, M. E., Lavin, D. H., and Asofsky, R. (1973) J. Natl. Cancer Inst. 51, 1509.
20. Gomard, E., Leclerc, J. C., and Levy, J. (1974) Nature, 250, 671.
21. Kiessling, R., Klein, E., Pross, H., and Wigzell, H. (1975) J. Immunol. 5, 117.
22. Sendo, F., Aoki, T., Boyse, E. A., and Buofo, C. K. (1975) J. Natl. Cancer Inst. 55, 603.
23. Herberman, R. B., Bartram, S., Haskill, J. S., Nunn, M. E., Holden, H. T., and West, W. H. (1977) J. Immunol. 119, 322.
24. Nelson, D. S., and Hopper, K. E. (1978) Cancer Letters, 5, 61.
25. Oliver, R. T. D. (1978) J. Royal Soc. Med. 71, 50.
26. Blum, H. F. (1959) Carcinogenesis by Ultraviolet Light, Princeton Univ. Press, Princeton, N. J.
27. Kripke, M. L. (1977) Cancer Res. 37, 1395.

28. Kripke, M. L. (1974) J. Natl. Cancer Inst. 53, 1333.
29. Kripke, M. L. and Fisher, M. S. (1976) J. Natl. Cancer Inst. 57, 211.
30. Fortner, G. W., and Kripke, M. L. (1977) J. Immunol. 118, 1483.
31. Lill, P. H., and Fortner, G. W. (1977) Proc. Am. Assoc. Cancer Res. 18, 55. (Abs.)
32. Fisher, M. S., and Kripke, M. L. (1977) Proc. Natl. Acad. Sci. USA, 74, 1688.
33. Daynes, R. A., and Spellman, C. W. (1977) Cell. Immunol. 31, 182.
34. Spellman, C. W., and Daynes, R. A. (1977) Transplantation, 24, 120.
35. Woodward, J. G., and Daynes, R. A. (1978) Cell. Immunol. 41, 304.
36. Daynes, R. A., Fernandez, P. A., and Woodward, J. G. (1979) Cell. Immunol. 45, 398.
37. Spellman, C. W., and Daynes, R. A. (1978) Cell. Immunol. 38, 25.
38. Daynes, R. A., Schmitt, M. K., Roberts, L. K., and Spellman, C. W. (1979) J. Immunol. 122, 2458.
39. Sell, S. (1978) Human Pathol. 9, 63.
40. Hook, R. R. Jr., Aultman, M. D., Adelstein, E. H., Oxenhandler, R. W., Millikan, L. E., and Middleton, C. C. (1979) Int. J. Cancer, 24, 668.
41. Oxenhandler, R. W., Adelstein, E. H., Haigh, J., Hook, R. R. Jr., and Clark, W. H. Jr. (1979) Am. J. Pathol. 96, 707.
42. Greene, M. H., and Fraremeni, J. F. Jr. (1979) in Human Malignant Melanoma, Clark, W. H. Jr., Goldman, L. I., and Mastrangelo, M. J. ed., Grune and Stratton, New York, pp. 139-166.
43. Hook, R. R. Jr., Aultman, M. D., Millikan, L. E., Oxenhandler, R. W., and Adelstein, E. H. (1977) Yale J. Biol. Med. 50, 561 (Abs.).
44. Berkelhammer, J., Caines, S. M., Dexter, D. L., Adelstein, E. H., Oxenhandler, R. W., and Hook, R. R. Jr. (1979) Cancer Res. 39, 4960.
45. Berkelhammer, J., and Hook, R. R. Jr. (1980) Transplantation (In Press).
46. Berkelhammer, J., Caines, S. M., Aultman, M. D., and Hook, R. R. Jr. (1979) Fed. Proc. 38, 1176 (Abs.).
47. Aultman, M. D., and Hook, R. R. Jr. (1979) Int. J. Cancer, 24, 673.

SUPPRESSOR CELLS AND TOLERANCE

RANDALL S. KRAKAUER[+] AND DIEGO SEGRE[++]
[+]Department of Rheumatic and Immunologic Disease, Cleveland
Clinic Foundation, Cleveland, Ohio, U.S.A. and [++]Professor
College of Veterinary Medicine, University of Illinois, Urbana,
Illinois, U.S.A.

Papers presented at the Suppressor Cell Workshop dealt with basic suppressor cell function and clinical immunology. Drs. Victoria Fraser and Helen Mullen of the Department of Medicine, University of Missouri, Columbia, Missouri, reported that mice given polyvinylpyrrolidone (PVP) coupled to spleen cells will three days later show a reduced response to PVP immunization in terms of IgM plaque forming cells. This tolerance is transferable to syngeneic animals and this transfer of tolerance is ablated by pretreatment before transfer of cells with anti-Thy-1. It appears therefore that suppressor cell-induced tolerance can be seen with this thymic independent antigen.

In studies of human suppressor cells, Drs. Nancy Goeken and John Thompson of the Department of Medicine, University of Iowa and VA Medical Center, Iowa City, Iowa, looked at the type of suppressor cell seen after allogeneic stimulation or incubation with the T cell mitogen Concanavalin A. The report coincides with previous reports that such cells suppress lymphocyte proliferation by fresh responding lymphocytes. However, they find that when assayed earlier than the optimal day (several days earlier), an enhanced proliferative response is seen, quantitatively equivalent to control untreated cultures. Thus, Concanavalin A-treated cells appear to cause a shift in the kinetics of cell proliferation. Both the early enhancing and late suppressive activities are mitomycin sensitive and presumably dependent upon proliferation and differentiation of precursors into enhancing or suppressor cells. Cells which are repeatedly exposed to alloantigen demonstrate suppressor activity, however, that is mitomycin resistant and not associated with early enhancement, presumably due to the pre-existence of mature suppressor cells rather than the generation of new ones. It has been suggested that two separate populations of suppressor cells are responsible for the mitomycin-sensitive and mitomycin-resistant effects. Alternately, the results may show the difference between suppressor cell precursors that are sensitive to mitomycin C and already existing mature suppressor cells not sensitive to mitomycin C.

Drs. Rosenberg and Lysz of the Immunology Laboratory, Hutzel Hospital Surgical Unit, Wayne State University, Detroit, Michigan, evaluated the effects of glucocorticosteroids on mixed lymphocyte reactions. In equivalent anti-inflammatory concentrations, methylprednisolone, dexamethisone and hydrocortisone all produced significant suppression of lymphocyte proliferation. The minimal concentration of methylprednisolone required for the suppression was 0.25 µg/ml and higher concentrations were not more effective in producing the result. No suppressor cell mechanism for this inhibition was proposed.

The previous evidence had implicated reduction in suppressor cell function in the pathogenesis of autoimmunity, specifically with respect to systemic lupus erythematosus, and indeed, Drs. Thomas Alexander and Randall Krakauer of the Cleveland Clinic Foundation, Cleveland, Ohio, concur that suppressor T lymphocytes in this disease are defective. In studying suppression due to high allogeneic T:B cell ratios in an in vitro IgM synthesis system, however, they report the existence of a different suppressor cell in patients with SLE receiving glucocorticoids. Unlike the suppressor T cell seen in normals, this suppressor cell, although having SRBC receptors, is not ablated by 1000 rads of irradiation. However, it is ablated by 2500 rads of irradiation. These suppressor cells could either be activated by the glucocorticoids or expressed due to loss of immunoregulation caused by both steroids and SLE, and may be a mechanism by which corticosteroids are therapeutic in this disease. Drs. Fox, Gordon, Hetzel, O'Hara, and Dinst of the Division of Transplantation, Department of Surgery, Henry Ford Hospital, Detroit, Michigan, induced suppressor cells by incubating the murine splenocytes with Concanavalin A. Suppressor cells were injected into normal mice prior to skin allograft. Suppressor cells generated in this manner were able to prolong allograft survival. When recipient and allografted animals were pre-irradiated, this prolongation of allograft survival was not seen. This suggests, as has previously been reported, a suppressor mechanism whereby the Concanavalin A-induced suppressor cell is radiation-resistant; however, its target suppressor cell is radiosensitive.

T-CELL DIFFERENTIATION

DAVID A. CROUSE[+], REGINALD K. JORDAN[++] AND J.G. SHARP[+]
[+]Department of Anatomy, University of Nebraska Medical Center,
42nd and Dewey Avenue, Omaha, Nebraska, U.S.A. and [++]Department
of Anatomy, University of Newcastle-upon-Tyne, Newcastle, England

INTRODUCTION

Lymphocytes which result from the interaction of progenitor cells
with the thymus and/or thymic humoral factors, acquire a unique set
of surface antigens and functional attributes which together class
them as T-lymphocytes (T-cells). Such T-cells, however, are by no
means homogeneous and consist of many subpopulations or sets of cells
at different levels of maturation and with different functions. Several
excellent recent reviews have dealt with some aspects of this process
in the mouse[1,2] and human[3,4]. The purpose of this report is to summarize
current knowledge of T-lymphocyte differentiation with emphasis on
recent technical and conceptual innovations. Special reference will
be made to areas which were presented and discussed at the T-Cell
Differentiation Workshop of the 8th Annual Midwest Autumn Immunology
Conference.

MURINE PRE-THYMIC PROCESSES

Over the past two decades, many studies have demonstrated that all
mature lymphohematopoietic cells in the peripheral circulation arise
from a totipotential hematopoietic stem cell population[5]. The differentiation of myeloid cells (erythrocytes, granulocytes and platelets) has
been investigated in detail using specific clonal assays and, as a
result, sets of differentiating and maturing cells have been proposed
which consist of restricted progenitors through fully functional mature
cells[6]. Similar sets of cells, with the first element described as
a pre-thymic, T-restricted progenitor cell, have been proposed for
the T-cell lineage[7,8]. Such a restricted progenitor class has been
demonstrated in murine radiation chimera studies[9,10], as well as in
clinical studies which presented a restricted appearance of G6PD hemi-
zygosity in myeloid cell classes of chronic mylogenous leukemia or
polycythemia vera patients[11]. Thus T-cell sets appear to be derived

from T-restricted progenitors which migrate into the thymus in the
adult rather than directly from totipotential hematopoietic stem cells
per se.[5]

MURINE INTRA-THYMIC PROCESSES

Morphologically the period of intrathymic differentiation is characterized by a graded reduction in cell size from large through medium
to small lymphocytes[12]. During this time, extensive proliferation
is a classic feature yet there is still a lack of correlation between
such kinetic observations and the classes of cells involved. Most
functional studies have demonstrated at least two major populations
of small lymphocytes: cortical and medullary[13]. These populations
differ with respect to functional competence, steroid sensitivity,
radiation sensitivity, traffic patterns, and surface marker characteristics. Earlier studies supported the concept of a common maturational
pathway (i.e. cortical to medullary to peripheral lymphocytes). More
recent studies have proposed that both major intrathymic populations
may contribute directly to the peripheral T-cell pool[14], although Stutman
has demonstrated that the turnover of murine steroid resistant lymphocytes (medullary) is very slow[1].

In either case, traffic through an intact thymus appears to be essential
for the conversion of pre-thymic progenitors to their more differentiated
progeny found both within the thymus and in the periphery[1]. Most investigators favor the thymic epithelial cell as the specific component of
the intrathymic non-lymphoid stroma which provides the cellular basis
for further differentiation of pre-thymic progenitors into precursor
populations of the T-cell lineage[1-4,7,8,14-21]. Indeed, there have
been suggestions that the absence of such a cellular interaction, as
occurs in the nu/nu mouse or following thymectomy, leads to an alternative
pathway of differentiation for the pre-thymic progenitor cells which
gives rise to natural killer (NK) cells[22-24]. Since there are numerous
reports or speculations about chemotactic factors from thymic stroma
(epithelium?), which are specifically attractive to the pre-thymic
T-restricted progenitor cell[25-27], the elaboration of such humoral
agents by the developing thymus may be an important early step in the
establishment of normal T-cell differentiation pathways.

Although the mechanisms by which progenitors interact with the thymus
are not fully established, it is clear that they may involve direct
cell contact and/or the production and limited diffusion of locally

acting factors. These types of interaction are embodied in the concept
of a thymic microenvironment which functions in the control of T-cell
differentiation. In vitro studies suggest that the immune system may
depend heavily upon paracrine effectors. Unfortunately, the role of
such paracrine effectors is technically difficult to investigate and
evaluate in vivo. Stutman's in vivo studies on the immunological restoration of neonatally thymectomized (NTX) mice with free thymus grafts
showed that pre-thymic T-progenitors must interact directly with the
non-lymphoid stroma of the thymus for their further proliferation and
differentiation[1]. Reconstitution by humoral factors liberated from
a thymus in a diffusion chamber was effective only in the first 45
days following thymectomy and transplantation[1]. It was proposed that
only those cells which had already interacted with the thymus would
differentiate further under the influence of thymic humoral factors.
Such cells were termed post-thymic-precursors (PTP).

Studies presented by Cullan et al.,[28] compared such free thymus
grafts or thymus grafts in diffusion chambers to thymic humoral factors
(thymosin, thymopoietin) in the restoration of immune and hematopoietic
function of NTX mice. Free thymus was better than thymus in a diffusion
chamber which in turn was slightly better than the thymic factors in
effecting reconstitution and the earlier reconstitution was performed
after thymectomy the better the extent of reconstitution. Reconstitution
at 90 days post-thymectomy was minimal even with free thymus grafts
leading to speculation that by this time either the pre-thymic precursor
pool had declined in efficacy or had been diverted elsewhere (NK cells?).
The concept of a PTP maturational step was generally supported by these
studies. It is interesting to note that there was full restoration
of specific immune responsiveness (GVH) when compared to minimal restoration of non-specific mitogen responsiveness. Analysis of other experiments involving reconstitution with thymic grafts demonstrates that
reconstitution, although significant when compared to non-reconstituted
animals, may be minimal when compared to age-matched intact controls
and often a selective reconstitution is observed. These observations
led to discussion as to the nature of thymic involvement in the differentiation of clones of T-restricted progenitor cells in this system. Although
the data is currently very limited it suggests that differentiation
may be hierarchical with selectivity biased towards alloreactive cells.
This, in turn, led to speculation that the thymus selects and amplifies

sets of pre-existing clones of germ-line determined specificities represented by the pre-thymic (and NK?) precursor cells[42]. Indeed, very recent reports[29] have pointed out the apparent development of a portion of the T-cell repertoire prior to any interaction of the pre-thymic cell with the thymic stroma. These studies also suggest the possibility that graft age and the timing of thymus-grafting of adult thymectomized allogeneic radiation chimeras may be important considerations in making comparisons to similar chimeras with an intact thymus.

Recognition necessary for antigen-specific responses of T-cells to virus-infected and chemically modified cells is restricted to self-MHC (major histocompatibility antigens). A major role of the thymus in T-cell differentiation is the selection of MHC restriction. Zinkernagel and co-workers[30,32] have shown that thymus restriction for anti-viral responses in radiation chimeras demonstrates; (1) specificity for self-MHC as expressed in the radioresistant portion of the thymus and independence of antigen, (2) full maturation of the response dependent on some lymphohematopoietic cells and the thymus being MHC compatible both for I and K or D, (3) some overlap of specificities between certain H-2 haplotypes, (4) possibly a requirement for T helper cells in the generation of cytotoxic effector cells. It is generally accepted that suppressor cells are not involved in the maintenance of the restricted repertoire of functionally competent radiation chimera T-cell populations[30,31,89,90]. Rather a mechanism involving either selective clonal expansion or selective clonal deletion of potential T-cell clones is implicated[89]. This does not exclude a role for haplotype specific suppression at the level of the thymus during T-cell differentiation and, in fact, such a process has been proposed[91]. Although thymic restriction of both virus and hapten-mediated cell responsiveness has been demonstrated in radiation chimeras, contradictory results have been reported for hapten-modified cell responses in neonatally tolerant mice and when responder cells were negatively selected for allogeneic reactivity (see 30,90). Recently, triparental chimeras capable of generating a cytotoxic response to hapten-modified cells of both marrow donor and host, but not to unrelated haplotypes have been described[90]. This suggests a role for the host environment in defining the T-cell repertoire but adds further evidence suggesting that restriction by the thymus may not be absolute.

In addition to in *vivo* studies, information on the interaction of

progenitor populations with the thymic stroma has been sought from
in vitro studies. These generally employ the principle of combining
selected target T-progenitors with thymus derived non-lymphoid cell
monolayers as a source of thymic factors or inducers in co-culture
protocols[7,15-21,33-37] and evaluating various endpoints. Such studies
have been conducted using a wide variety of target populations and
thymic "epithelium" from several species including man. Although the
induction of markers characteristic of mature T-cells and the promotion
of limited function have been achieved by such in vitro techniques,
full restoration of mature function has yet to be achieved. Sharp
et al.,[38] presented data on correlated morphological and functional
studies for several species, showing similar restoration effects which
were only partially dependent upon the apparent presence of morphologically appearing epithelial cell populations in the monolayers.
The heterogeneity of cell types in such in vitro systems, particularly
in the mouse, has previously been reported[39] with parallel in vivo
transplantation studies which clearly demonstrated the presence of
potentially functional thymic epithelium[40].

Considering the problems inherent in such monolayer systems, other
in vitro protocols initiated with fetal thymus (14 day) organ cultures
have been described[41]. In these studies it was shown that the use
of low temperature organ culture ($24^{\circ}C$-7 days) permits the development
of purified thymic epithelium for transplantation studies. Maintenance
of epithelium was much better in organ cultures prepared by this technique
than in thymic explants. Monolayer preparations survived very poorly
in this protocol, suggesting that three-dimensional spatial integrity
of the thymus is important to the survival and maintenance of thymic
epithelium. The reasons for this are not known, although speculation
was offered that perhaps neural and/or neuroendocrine factors might
be important in the process[38]. The organ culture system will now allow
a much more critical evaluation of the role of thymic epithelial cells
in the processes of pre-thymic (chemotaxis), intra-thymic (MHC restriction)
and post-thymic (humoral factors) T-cell differentiation.

In addition to the role of the epithelial cell population of the
thymic stroma, there is now considerable evidence that thymic macrophages
may be intimately involved in the T-lymphocyte differentiation process.
Some evidence comes from somewhat peripheral studies which have simply
demonstrated large numbers of macrophages[43,44] and their precursors[45]

in the thymus as well as studies which have demonstrated in vitro a
non-random reassociation or rosette formation between thymic lymphocytes
and macrophages[7,26,46-48]. Some evidence[49,50] also concerns certain
humoral products of thymic macrophages which appear to play a role
in the generation of various differentiation markers and mature function
in thymic lymphocytes. These factors have been shown to be different
from the classical thymic humoral factors and may be part of an overall
multisignal differentiation system. Thymic macrophages may represent
an important component of the cooperating lymphohematopoietic cell
population required for complete expression of T-cell function. These
cells may play a significant role in the expansion and maturation of
PTP and could impose a degree (quantitative rather than qualitative?)
of selectivity on immune responsiveness at the PTP stage[30]. The characteristics (ontogeny, kinetics, radiosensitivity, migration patterns)
of these thymic macrophages in comparison to other antigen presenting
cells needs to be defined.

By far the great majority of studies concerning humoral promotion
of differentiation in the T-lymphocyte set have focused on purportedly
epithelially-derived circulating thymic humoral factors. Studies with
these factors have been extensively evaluated in several reviews[51-53],
thus only a brief summary will be presented here. It is clear that
all the factors are protein or polypeptide in nature with a typical
molecular weight of less than 40,000. Two of the factors (thymopoietin
and serum thymic factor) have been sequenced and synthesized, yet bear
no striking amino acid homologies in spite of their reported similar
functions[53]. Thymosin fraction V has been fractioned into about 20
components (α, β and γ thymosins) which have been reported to promote
differentiation of different markers or subclasses of T-cells[54]. By
far the most active of these components is designated α, which is 10-
1000 times more active than thymosin fraction V in several assay systems[54].
Since most of the other factors remain relatively uncharacterized bio-
chemically, their relationship in terms of active functional sites
and amino acid sequence, is still uncertain. All of the factors appear
not to be species specific since the bovine, swine and human derived
materials have been shown to induce some level of differentiation in
several different target models[51-59]. It should be pointed out that
most of the "induction of differentiation" was quantitated by the generation of mature surface markers or binding characteristics associated

with more mature populations. The induction of functional maturity
has only been reported in a few cases and then not to levels associated
with a normal intact lymphoid system. Some experiments have clearly
demonstrated an inability of thymic humoral factors to restore immune
function completely[60,61]. Two of the humoral factors (thymosin and
thymopoietin) were reported not to provide any reconstitution of GVH
activity in 90 day old NTX mice[28]. Since many workers now feel that
the role of a thymic hormone probably resides in the late stages of
differentiation which occur mainly in the periphery[1,7,15,62], it is
not surprising that 90 day old NTX mice, whose post-thymic precursors
were at very low levels, were not reconstituted by thymic factors.
Finally, there is good evidence to suggest that such thymic factors
may act by elevating intracellular levels of cyclic AMP[63-65] and/or
cyclic GMP[66-67]. These experiments suggest that the cAMP signal may
act on an early stage of T-cell differentiation while cGMP may act
at a later stage or simply to expand a differentiated T-cell population.
Isoproterenol, a stimulator of cAMP, was found to have effects similar
to those observed using thymic factors[28].

Intrathymic processes, therefore, include differentiation mediated
by cell contact as well as by humoral factors. The resulting cell
population(s), upon migration to the periphery, may undergo further
differentiation under the influence of the same or different humoral
factors, and eventually provide the complete repertoire of fully mature
sub-sets of T-cells which function at various levels in the many aspects
of the T-lymphocyte arm in the immune system. The actual differentiative
link between the post-thymic precursor and fully mature T-cell subpopulations is still one of the least understood areas of the entire
T-cell differentiation process. The mechanisms by which functional
sets are derived and their relationship to common versus parallel T-
progenitor/precursor series are yet unclear and require considerable
further research.

HUMAN PRE-THYMIC AND INTRATHYMIC PROCESSES

Most of the previous discussion has been derived from studies in
the mouse, yet there is increasing experimental and clinical evidence
which suggests that the T-cell differentiation process in the human
is very similar. The differentiation and maturation of human T-lymphocytes has been reviewed in considerable detail in several recent
publications.[3,4,53,68,69]

The obvious differences in T-lymphocyte differentiation of mouse
and man are closely tied to their ontogenetic development, the full-
term human fetus being much more immunologically mature than the full-
term mouse fetus[68]. Obviously, experimental manipulation is not possible
in the human in vivo system but an analysis of congenital immunodeficiency
diseases (IDD) and their treatment has provided considerable insight.
Evaluation of various classes of IDD patients treated by administration
of thymic humoral factors[51,53,70-74] or thymic transplantation[4,53,69,75]
has demonstrated that the initial step of T-cell differentiation in
man probably requires contact with epithelial cells while subsequent
steps may be driven by humoral factors[4]. Thus it appears that distinct
stages of human T-cell differentiation parallel those described by
Stutman[1] as pre-thymic precursor, post-thymic precursor and mature
functional populations. As in the murine system, one of the current
areas of most active study centers on the definition of peripheral
T-cell sub-sets and attempts to clarify their precursor populations
as well as maturational pathways[82-84].

MURINE POST-THYMIC PROCESSES

Although murine peripheral T-lymphocytes are cells with rather uniform
morphology, the population is quite heterogeneous and can be shown
to be composed of many sub-sets of cells. Initially these cells were
segregated purely on the basis of function in various in vivo or in
vitro assays. Populations of T-cells involved in suppressor, helper
or cytotoxic activities were clearly established[76,77]. Many recent
studies have been directed at determining the differentiation pathways
which lead to the functional populations. Most of these investigations
have relied on the detection of specific surface markers which are
characteristic of some T-cell classes. Although many different T-cell
specific surface proteins have been examined, including a new set of
three high molecular weight (173K, 187K and 200K Daltons) proteins
described at this workshop[78], the bulk of experimental data, directly
related to functional subpopulations, is based on the Lyt series of
alloantigens[76,79].

At least three separate sub-sets of T-cells have been identified
in the peripheral murine lymphocyte pool. The sub-set expressing
Lyt 1+,2+,3+ is the largest and makes up about 50% of the peripheral

T-cell pool[76]. Stutman[1] has shown that the post-thymic T-precursor
is of this phenotype and thus at least some of the Lyt 1+,2+,3+ cells
are sensitive to the influence of thymic humoral factors and are not
fully mature. The differentiation of some of the Lyt 1+,2+,3+ cells
can lead to the production of both Lyt 1+ and Lyt 2+,3+ populations.
Cells of the same set (Lyt 1+,2+,3+) have thus been described as "regulatory"
cells[76] which regulate the supply of more mature sets by directly giving
rise to the cells or possibly by exerting other immunoregulatory effects.
A second sub-set described as "inducer" cells[76] comprising 30% of the
peripheral T-cells is of the Lyt 1+ phenotype. Such cells "help" B-
cells in the generation of an antibody response as well as acting on
macrophages to invoke their participation in delayed-type hypersensitivity
responses. They may also induce the precursors of cytotoxic cells
to complete their differentiation and become killer-effectors, and
may be involved in the induction of a feedback inhibition pathway[76].
More recent evidence suggests that some of the Lyt 1+ cells may also
be involved in inductive influences upon other sets of cells outside
the lymphoid class, i.e. osteoclasts and erythroid precursors[76]. The
third class of T-cells identified by these means are the Lyt 2+,3+
cytotoxic and suppressor cells which only make up 5-10% of the peripheral
T-cell pool. It is not known at this time if one cell can provide
both functions or if they reside in different sub-classes of Lyt 2+,3+
cells. Some recent evidence[80] seems to indicate that the populations
are separable. The suppressive activity of the Lyt 2+,3+ cells which
is active on both humoral and cell-mediated responses, is not mediated
by a cytotoxic mechanism. Such a suppressive action (not correlated
to Lyt phenotype) was demonstrated to be independent of cytotoxic activity
directed at either the responder or stimulator population in a CML
assay[81]. Further separation of these T-cell sets based on the Qa-1,
Tla, Ia, other Ly and surface markers have been proposed by numerous
workers and are described in recent reviews[76,79].

HUMAN POST-THYMIC PROCESSES

   In the human, very similar sets of T-cell subpopulations have been
described but based on quite different surface markers[82-85]. Receptors
for immunoglobulins (Ig) have been demonstrated on some T-cells as
well as other lymphoid cell types. Most T-cells were found to have
IgG receptors (Tγ) while smaller percentages displayed IgM (Tμ), or

IgA ($T\alpha$) receptors[84,85]. In all cases the specificity of the receptor
appears to be for the Fc portion of the immunoglobulin molecule[84].
It has been found that the $T\gamma$ cells comprise about 50% of the peripheral
T-cells and are most active in a helper function for B-cell differentiation
in response to antigen[84]. However, some T-cells appear to act in a
suppressor fashion or under some circumstances be induced to switch
to cells with receptors only for IgM[84]. Thus the functional classification
of these cells is not clear at this time. The $T\mu$ population comprises
about 10% of the peripheral pool and has been classified as a suppressor
population[82,84]. Again these cells do not effect suppression by a
cytotoxic mechanism and are required at the very early stages of antigen
stimulated B-cell differentiation to exert their regulatory function[82].
Unfortunately, the ontogeny of these subpopulations in the human is
simply not known and techniques by which such peripheral T-cells can
be evaluated are an active area of research. A number of culture systems
which use conditioned medium[86] or feeder layers[87] have been described
which allow the long term culture of primed or cloned human T-cells
with full retention of function and specificity. Similar in vitro
systems applied to precursor populations or combined with agents which
are known to influence T-cell differentiation have been used on selected
T-cell populations to examine T-cell differentiation pathways in the
human[88]. Although the results at this time are inconclusive, such
techniques continue to support a valuable avenue of research.

CONCLUSION

This brief summary of T-cell differentiation provides an overview
of the current literature with emphasis on the workshop topics and
points out some of the major questions yet unanswered in this area
of study. It is clear that T-cell differentiation in mouse and man
are essentially similar. Both processes begin with the hematopoietic
stem cell which in turn gives rise to a T-restricted pre-thymic progenitor
population. Under chemotactic influence, these cells migrate into
the thymus where they undergo extensive proliferation and differentiation
which is influenced by the thymic microenvironment to establish sets
of precursor cells specific for the T lineage. The thymus plays a
major role in the development of the T cell repertoire of radiation
chimeras in that recognition of virus-infected or hapten-modified cells
are restricted to self-MHC by the thymus. Additionally I and K or D

identical co-operating lymphohematopoietic cells are required for full maturation of responsiveness. Evidence from other systems suggests that thymic restriction of responsiveness may not be absolute. Thymic macrophages may play an important role in the expansion of post-thymic precursor cells into sets of maturing T-cell subpopulations. These maturing cells migrate to the periphery and undergo a final maturation under the influence of thymic humoral factors. This process may also be regulated in part by other populations of differentiating T-cells. The end-product of this differentiation network is a functionally hetero-geneous population of morphologically similar T-cells which provide all the necessary interactions and terminal effects necessary to maintain immunological homeostasis.

ACKNOWLEDGMENTS

The authors would like to thank all of the workshop registrants for their cooperation and enthusiastic participation in both the poster and platform sessions. Likewise, we would like to thank Drs. A.L. Goldstein for the thymosin fraction V and G. Goldstein for the thymopoietin (TP5) preparations used in studies presented at the workshop. Our thanks are due also to Mrs. Sally Mann and Mrs. Roberta Anderson for help in the preparation and typing of the manuscript. (The authors' work reported in this paper was supported by NIH Grants AI 15819, AM21137 and UNMC Seed Grant Funds.)

REFERENCES

1. Stutman, O. (1978) Immunological Rev., 42, 138.
2. Owen, J.J.T., Jordan, R.K., Robinson, J.H., Singh, U. and Wilcox, H.N.A. (1977) Cold Spring Harbor Symp., 41, 129.
3. Touraine, J-L. (1978) in The Pharmacology of Immunoregulation, G.H. Werner and F. Floc'h, eds., Academic Press, New York, p.63.
4. Gelfand, E.W., Dosch, H.M. and Shore, A. (1978) in Hematopoietic Cell Differentiation, D.W. Golde, M.J. Cline, D. Metcalf and C.F. Fox, eds., Adademic Press, New York, p. 277.
5. Metcalf, D. (1977) Recent Results in Cancer Res., 61, 1.
6. Lajtha, L.G. (1979) Differentiation, 14, 23.
7. Jordan, R.K., Crouse, D.A., Harper, C.M., Watkins, E.B. and Sharp, J.G. (1979) in Experimental Hematology Today 1979, S. Baum and G.D. Ledney, eds., Springer-Verlag, New York, p. 139.
8. Gorczynski, R.M. and MacRae, S. (1977) Immunology, 33, 697.
9. Abramson, S., Miller, R.G. and Phillips, R.A. (1977) J. Exp. Med., 145, 1567.
10. Basch, R.S. and Kadish, J.L. (1977) J. Exp. Med., 145, 405.
11. Failkow, P.J. (1974) N. Eng. J. Med., 291, 26.
12. Sainte-Marie, G. and Leblond, C.P. (1964) in The Thymus in Immunobiology, R.A. Good and A. Gabrielsen, eds., Harper and Row, New York, p.207.

13. Owen, J.J.T. and Raff, M. (1970) J. Exp. Med., 132, 1216.
14. Goldschneider, I. (1976) Cell. Immunol., 24, 289.
15. Kruisbeek, A. (1979) Thymus, 1, 163.
16. Mosier, D.E. and Pierce, C.W. (1972) J. Exp. Med., 136, 1484.
17. Willis, J.I. and St. Pierre, R.L. (1976) Adv. Exp. Biol. Med., 73A, 111.
18. Waksal, S.D., Cohen, I.R., Waksal, H.W., Wekerle, H., St. Pierre, R.L. and Feldman, M. (1975) Ann. N. Y. Acad. Sci., 249, 492.
19. Papiernik, M., Nabarra, B. and Bach, J.F. (1975) Clin. Exp. Immunol., 19, 281.
20. Pyke, K.W. and Gelfand, E.W. (1974) Nature (London), 251, 421.
21. Loor, F. (1979) Immunology, 37, 157.
22. Heberman, R.B., Djeu, J.Y., Kay, H.D., Ortaldo, J.R., Riccardi, C., Bonnard, G.D., Holden, H.T., Fagnani, R., Santoni, A. and Puccelli, P. (1979) Immunol. Rev., 44, 43.
23. Cantor, H., Kasai, M., Shen, F.W., LeClerc, J.C. and Glimcher, L. (1979) Immunol. Rev., 44, 3.
24. Sharp, J.G. and Crouse, D.A. (1980) in Experimental Hematology Today 1980, S. Baum and G.D. Ledney, eds., Springer-Verlag, New York (in press).
25. Kindred, B. (1978) Immunol. Rev., 42, 60.
26. Schulte-Wissermann, H., Borzy, M.S., Albrecht, R. and Hong, R. (1979) Scand. J. Immunol., 9, 45.
27. Pyke, K.W., Papiernik, M. and Bach, J.F. (1979) J. Immunol., 123, 2316.
28. Cullan, G.M., Harper, C.M., Anderson, R.A., Sharp, J.G. and Crouse, D.A., this workshop.
29. Cohn, M.L. and Scott, D.W. (1979) J. Immunol., 123, 2083.
30. Zinkernagel, R.M. and Doherty, P.C. (1979) Adv. Immunol., 27, 51.
31. Bevan, M.J. and Fink, P.J. (1978) Immunol. Rev., 42, 3.
32. Zinkernagel, R.M., Callahan, G.N., Althage, A., Cooper, S., Klein, P.A. and Klein, J. (1978) J. Exp. Med., 147, 882.
33. Gorczynski, R.M. and MacRae, S. (1979) Immunology, 38, 1.
34. Hensen, E.J., Hoefsmit, E.C.M. and Van den Tweel, J.G. (1978) Clin. Exp. Immunol., 32, 309.
35. Goust, J.M., Vesole, D.H. and Fudenberg, H.H. (1979) Clin. Exp. Immunol., 38, 348.
36. Sato, V.L., Waksal, S.D. and Herzenberg, L.A. (1976) Cell. Immunol., 24, 173.
37. Gershwin, M.E., Ikeda, R.M., Kruse, W.L., Wilson, F., Shifrine, M. and Spangler, W. (1978) J. Immunol., 120, 971.
38. Sharp, J.G., Crouse, D.A., Jordan, R.K., Grazulewicz, S. and Hoppe, L., this workshop.
39. Jordan, R.K. and Crouse, D.A. (1979) J. Reticuloendothel. Soc., 26, 389.
40. Jordan, R.K., Crouse, D.A. and Owen, J.J.T. (1979) J. Reticuloendothel. Soc., 26, 373.
41. Jordan, R.K. and Crouse, D.A. (1980) in Development and Differentiation of Vertebrate Lymphocytes, J.D. Horton and J.B. Solomon, eds., Elsevier/North-Holland, New York (in press).
42. Kaplan, J., Sharp, J.G., discussion this workshop.
43. Kostowiecki, M. (1963) Z. mikr.-anat. Forsch, 69, 585.
44. Bearman, R.M., Levine, G.D. and Bensch, K.G. (1978) Anat. Rec., 190, 755.
45. MacVittie, T.J. and Weatherly, T.L. (1977) in Experimental Hematology Today, S. Baum and G.D. Ledney, eds., Springer-Verlag, New York, p. 147.

46. Sharp, J.A. (1971) J. Path., 103, 87.
47. Siegel, I. (1970) J. Immunol., 105, 879.
48. Lipsky, P.E. and Rosenthal, A.S. (1973) J. Exp. Med., 138, 900.
49. Beller, D.I., Farr, A.G. and Unanue, E.R. (1978) Fed. Proc. (US), 37, 91.
50. Van den Tweel, J.G. and Walker, W.S. (1977) Immunology, 33,817.
51. Bach, J.F. and Carnaud, C. (1976) Prog. Allergy, 21, 342.
52. Van Bekkum, D.W. (1975) The Biological Activity of Thymic Hormones, Kooyker Sci. Pub., Rotterdam.
53. Phawa, P., Ikehara, S., Phawa, S.G. and Good, R.A. (1979) Thymus, 1, 27.
54. Goldstein, A.L., Low, T.L.K. and Thurman, G.B. (1980) Ann. N. Y. Acad. Sci., (in press).
55. Pazmino, N.H., Khle, J.N. and Goldstein, A.L. (1978) J. Exp. Med., 147, 708.
56. Rotter, V. and Trainin, N. (1979) J. Immunol., 122, 414.
57. Astaldi, G.C.B., Astaldi, A., Groenewoud, M., Wijermans, P., Schellekens, P.T.A. and Eijsvoogel, V.P. (1977) Eur. J. Immunol., 7, 836.
58. Goldstein, G., Scheid, M.P., Boyse, E.A., Schlesinger, D.H. and Van Wauwe, J. (1979) Science, 204, 1309.
59. Bach, J.F., Dardenne, M., Pleau, J.M. and Bach, M.A. (1975) Ann. N. Y. Acad. Sci., 249, 186.
60. Primus, J.F., DeMartino, L., MacDonald, R. and Hansen, H.J. (1978) Cell. Immunol., 35, 25.
61. Martinez, D., Field, A.K., Schwam, H., Tytell, A.A. and Hilleman, M.R. (1978) Proc. Soc. Exp. Biol. Med. (U.S.), 159, 195.
62. Yakir, Y., Kook, A.I. and Trainin, N. (1978) J. Exp. Med., 148,71.
63. Horowitz, S.D. and Goldstein, A.L. (1978) Clin. Immunol. Immunopath., 9, 408.
64. Bach, M.A., Fournier, C. and Bach, J.F. (1975) Ann. N. Y. Acad. Sci., 249, 316.
65. Scheid, M.P., Goldstein, G., Hammerling, U. and Boyse, E.A. (1975) Ann. N. Y. Acad. Sci., 249, 531.
66. Naylor, P.H., Sheppard, J., Thurman, G.B. and Goldstein, A.L. (1976) Biochem. Biophys. Res. Commun., 73, 843.
67. Sunshine, G.H., Basch, R.S., Goffey, R.G., Cohen, K.W., Goldstein, G. and Hadden, J.W. (1978) J. Immunol., 120, 1594.
68. Stutman, O. and Caulkins, C.E. (1977) Handbuch der allgemeinen Pathologie VI/8, 169.
69. Gelfand, E.W. and Dosch, H-M. (1979) in Regulation by T-Cells, D.G. Kilburn, J.G. Levy and H-S. Teh, eds., Univ. of British Columbia, Vancouver, p. 246.
70. Touraine, J-L., Incefy, G.S., Touraine, F., L'Espérance, P., Siegal, F.P. and Good, R.A. (1974) Clin. Immunol. Immunopath., 3, 228.
71. Incefy, G.S., Grimes, E., Kagan, W.A., Goldstein, G., Smithwick, E., O'Reilly, R. and Good, R.A. (1976) Clin. Exp. Immunol., 25,462.
72. Gupta, S., Kapoor, N., Goldstein, G. and Good, R.A. (1979) Clin. Immunol. Immunopath., 12, 404.
73. Wara, D.W. and Ammann, A.J. (1978) Transplant. Proc., 10, 203.
74. Goldstein, A.L., Thurman, G.B., Low, T.K.L., Rossio, J.L. and Trivers, G.E. (1978) J. Reticuloendothel. Soc., 23, 253.
75. Hong, R., Schulte-Wisserman, H., Horowitz, S., Borzy, M. and Findlay, J. (1978) Transplant. Proc. (U.S.), 10, 201.
76. Cantor, H. and Gershon, R.K. (1979) Fed. Proc. (U.S.), 38, 2058.
77. Katz, D.H. and Benacerraf, B. (1972) Adv. Immunol., 15, 1.
78. Dunlap, B.E., Mixter, P., Watson, A. and Bach, F.H., this workshop.
79. McKenzie, I.F.C. and Potter, T. (1979) Adv. Immunol., 27, 179.

80. Murphy, D.B. (1978) Springer Seminar Immunopathol., 1, 111.
81. Orosz, C. and Bach, F., this workshop.
82. Moretta, L., Ferrarini, M. and Cooper, M. (1978) Contemp. Top. Immunobiol., 8, 19.
83. Chess, L. and Schlossman, S.F. (1977) Contemp. Top. Immunobiol., 7, 363.
84. Gupta, S. and Good, R.A. (1979) Thymus, 1, 135.
85. Lum, L.G., Muchmore, A.V., Karen, D., Vecker, J., Koski, I., Strober, W. and Blaese, R.M. (1979) J. Immunol., 122, 65.
86. Klassen, L.W., Brown, S., Thompson, J.S. and Goeken, N., this workshop.
87. Hank, J., Inouye, H., Guy, L., Alter, B.J. and Bach, F., this workshop.
88. Gupta, S. (1979) J. Immunol., 123, 2664.
89. Hodes, R.J., Hatchcock, K.S. and Singer, A. (1980) J. Immunol., 124, 134.
90. Lattine, E.C., Gershon, H.E. and Stutman, O. (1980) J. Immunol., 124, 274.
91. Miller, R.G. (1979) in Regulation by T-cells, D.G. Kilburn, J.G. Levy and H-S. Teh, eds., Univ. of British Columbia, Vancouver, p. 77.

SECRETORY IMMUNITY AND IMMUNOGLOBULIN TRANSPORT

J. E. BUTLER

Department of Microbiology, University of Iowa, Iowa City, Iowa 52242

The presentation and discussion of six posters plus oral presentation by
three additional participants were concentrated in two major areas: (1) Those
concerning the secretory and systemic antibodies associated with experimental
studies of the lung and (2) those concerned directly or indirectly with the
transport of immunoglobulins.

Data were presented by W. D. Geoghegan et al. which showed that serum IgG
antibodies and their F(ab')2 and Fab fragments could prevent the uptake of
aerosally administered ovalbumin (OA). The non-Fc-dependent effector mechanism
by which these antibodies mediate this blockage was unexplained. Whether this
blocking IgG (or its fragments) actually enter the alveolar fluid was not
known. It was suggested by Calvanico that the protective humoral antibodies
in the distal portion of the lung might be of a different isotype, i.e., IgG,
than those effecting protection in the proximal, hylar region, i..e., IgA thus
suggesting roles for both antibody classes in the lower respiratory tract.
Swanson et al. reported on immunohistochemical studies which showed that both
IgG and IgA are present in the lungs of normal rabbits and healthy rabbits.
"Healthy" rabbits are those that had received repeated aerosal challenges of
OA and that had recovered from the acute alveolitis associated with a single
aerosol challenge in parenterally primed animals. It was shown using
double-labelled immunofluorescence that the lungs of recovered animals also
contained anti-OA; IgA anti-OA was associated preferentially with alveolar
macrophages and alveolar lining while IgG anti-OA was localized in plasma
cells as well as lining the alveoli. It was also shown histologically by
Swanson et al that animals suffering from alveolitis showed a loss of the
immunoglobulin layer which covers the alveolar cells of their healthy counterparts. The latter observation might be correlated with the observations
of Calvanico who reported that during the time-course of alveolitis, IgA is
released into the lavage fluids of the lung to such an extent that the levels
of IgA > IgG. It was not known whether this increase in IgA was also IgA-

antibody to the disease-causing antigen. The possible roles of macrophages in the surface adsorption and intracellular uptake of IgA were briefly discussed.

The humoral response in the lung therefore appears to be complex and it would seem incorrect to believe the dogma that local IgA is the only important antibody. In fact, IgA-positive cells producing specific antibodies are difficult to find (Swanson et al.) suggesting that the IgA found in the lung may arise from IgA-producing cells found elsewhere.

The possibility that local IgA could result from remote site antigenic stimulation and subsequent transport though serum were issues that received considerable theoretical discussion and some experimental support. Halsey et al. using myeloma immunoglobulins, showed that suckling neonatal mice absorb into their serum preferentially iodinated, oligomeric IgA and IgM as opposed to IgG when these labelled immunoglobulins are administered to the lactating mother. Human oligomeric IgM was absorbed in the same manner as homologous IgM. These interesting data, while suggesting selective transfer of those immunoglobulin which can bind secretory component, were somewhat compromised by two issues. First, the surprising transfer of heterologous albumin and secondly, the argument that transport to the mammary gland might be a highly specialized physiological process that is not necessarily universal for all secretory epithelia. Halsey also discussed the steady-state equilibrium which could explain how large quantities of IgA could be transported out of the serum while at the same time maintaining very low levels of this immuno-globulin in serum.

As many studies of immunoglobulin transport are performed by intravenous injection of iodinated immunoglobulins, the purity of these passively administered proteins is critical to the eventual interpretation of the data. Obtaining pure immunoglobulins may be difficult in species where myeloma proteins are not available. An example of this problem was described by McGivern et al. who showed that dimeric bovine IgA in colostrum cannot be readily separated from dimeric IgG1 in the same body fluid. Bovine IgA and IgG1 dimers possess overlapping ionic and molecular size characteristics. Even in the presence of 1% SDS or 1 M NaCl some IgG1 dimers are not deaggre-gated. While most contaminating dimeric IgG1 can be removed using a S. aureus Protein-A affinity column, highly purified secretory IgA can be readily prepared from broncho-alveolar washings in which dimeric IgG1 is absent. Questions pertaining to the nature of the SDS-resistent IgG1 dimers were posed, although data on their chemistry were not available.

Secretory component (SC) has been implicated as playing a major role in the transport of oligomeric IgA and IgM. Schiff mentioned the importance of studying the binding of SC to these molecules and pointed out that kinetic studies which indicated a higher intrinsic affinity of SC for IgM than for dimeric IgA need reinterpretation. The problem of the binding of SC and its role in transport is further complicated in rats and mice by the occurrence of two antigenically distinct SCs; "slow" and "fast". Koertge et al. reported on data collected using sucrose density gradient ultracentrifugation and the ELISA-based Antigen Distribution Assay to study slow and fast SC. These data indicated that rat IgA is polymeric in all body fluids and that this IgA in bile, intestinal washings, saliva and lung washings, contains both "slow" and "fast" SCs while thoracic duct lymph and serum IgA probably contain neither. The data also show that "slow" rat SC is especially abundant in rat saliva. Whether both SCs are present on the same or different IgA molecules was not determined. While the role of these dual SCs is not clear, the number of theories to explain the function of SCs still out-numbers the number of different SCs described.

The possible roles of SCs were further discussed. While there is support for the theory that SC serves as the receptor on epithelial cells which recognizes oligomeric IgA (or IgM) and hence facilitates transport, the reason for the retention of the SC after transport remains an enigma. While it is unusual for most transport receptors to be transported with the molecules they transport, Schiff pointed out that the extremely high binding efficiency of SC for IgA may be the common basis for the highly efficient removal of IgA from serum as well as the failure for SC to dissociate as do other transport proteins. The reason why a covalent SC-IgA bond forms during this transport (which would seemingly further hinder dissociation) remains unexplained.

The issue of heterogeneity of rabbit SC remains unresolved. Preliminary studies of affinity purified rabbit SC suggested that it may show a "slow" component similar to that of the rat. In lieu of the occurrence of enzyme degradation products of SC which also migrate in polyacrylamide gel electrophoresis similar to slow rat SC, such preliminary data must be cautiously regarded. Hanly pointed out that one potentially valuable approach to the probelm of heterogeneity of rabbit SC would involve extension of their studies of rabbit SC allotypes.

Finally, Butler et al. presented data to indicate that during the uptake of maternal colostral immunoglobulins by neonatal piglets, IgG is exclusively absorbed by the villus epithelial cells of the piglet gut while cells in the crypt appear to specifically adsorb IgM and IgA; possible in the case of the latter, by specific interaction with SC. The exact identity of the crypt cells which adsorb IgA and IgM was not established. Such data suggest that care must be used in the general interpretation of epithelial cell transport activity in secretory immunity because columnar epithelial cells can function differentially in either secretory or absorptive activity.

The workshop was well-attended and served to indicate that at least in the two areas of secretory immunity covered by the work of the participants, the future promises some exciting challenges for scientists and students.

INFECTION OR ADJUVANT-INDUCED ALTERATION OF THE IMMUNE RESPONSE

WARD E.BULLOCK
Department of Medicine, University of Kentucky College of Medicine, Lexington,
Kentucky 40536

The first annual workshop discussion on infection or adjuvant-induced alter-

ation of the immune response was initiated by E. Baron (co-worker R. Proctor)

from the University of Wisconsin who described a methodology designed to elicit

mouse peritoneal exudates that contain the largest possible number of polymor-

phonuclear (PMN) leukocytes relative to other leukocytes. Most successful was

the intraperitoneal injection of sterile 0.2% glycogen in a volume of 2.5 ml.

The exudate collected 4.5 h later contained 75-85% PMN cells. The use of

sterile 3.0% thioglycolate did not induce larger numbers of PMN cells. C57BL/6

mice gave higher yields of PMN cells than did BALB/c mice however large varia-

tions in cell yields were observed in both strains.

A summary of work in progress on the physical factors involved in adjuvantic-

ity was presented by F. Strickland (co-workers R. Pelley, R. Hunter, and N.

Schmolka) of the University of Chicago. Specifically, the relationship between

oil chain length and oil phase retention of antigen to in vivo antigen retention

and the humoral immune response was examined. Oil in water emulsions of ^{125}I-

bovine serum albumin (BSA) was prepared using a series of unbranched alkane

hydrocarbons. Increasing hydrocarbon chain length increased the stability of

the emulsion in vitro, increased antigen retention at the injection site, and

increased the antibody response to BSA.

To gain more flexibility over control of antigen retention and release, a

variety of detergents were incorporated into the oil and water emulsion. Incor-

poration into emulsions of certain pluronic polyols (non-ionic detergents) with

destabilizing effects such as L31, resulted in minimal antigen retention both in vitro and in vivo and produced a minimal humoral immune response to BSA.

Other pluronic polyols, for example L121, act to stabilize emulsions and these induced much greater antigen retention in vivo and improved the humoral immune response. However, it was noted that the oil phase stability of emulsions containing L121 were quite heterogeneous in vivo. Therefore, it was postulated that polyol/oil mixtures of this type, in addition to physically trapping antigen in the oil, may in some way alter the mechanisms of antigen processing to effect an increased humoral immune response.

In the next segment of the morning discussion, a group from the Wood Veterans Administration Medical Center in Milwaukee, Wisconsin, discussed their ongoing work using Bacillus Calmette-Guerin (BCG) to develop chronic pulmonary inflammation (CPI) in mice for study of the genetic factors and other mechanisms that modulate the inflammatory response in lung tissues. Those participating in the studies were V. Moore, D. Schrier, E. Allen, and J. Sternick. In initial experiments, the group observed that only certain strains of mice developed intense CPI and splenomegaly when injected intravenously with killed BCG; the capacity to mount this response is not controlled by genes in the H-2 complex. Evidence was presented to show that gene products of the Ig locus influence responsiveness by using BXD recombinant inbred mice and mice congenic at the Ig locus. It appears that at least three genes are involved in responsiveness as determined by breeding studies using F2 mice; the other loci are not known at the present time.

In other studies the Wood VA group has determined that the inability of CBA mice to develop CPI is at least partially explained by the fact that they possess cyclophosphamide (Cy)-sensitive suppressor T lymphocytes. Thus, administration of Cy to CBA mice permits the animals to express CPI that actually can be reversed by administering spleen cells from BCG-injected CBA mice not treated with Cy. Because the ability to respond is linked to the Ig complex, it was suggested that regulation of CPI and splenomegaly in CBA mice may involve feedback inhibition that might be related to VH receptors on T lymphocytes.

In mice that do develop CPI and splenomegaly (i.e. C57BL/6 mice), another form of splenic suppressor cell activity was detected that seems to be mediated by macrophage-like cells that act in nonspecific fashion; they suppress responses to mitogens, antibody synthesis, and delayed hypersensitivity. It was hypothesized that the macrophage-like cells modulate CPI and splenomegaly by releasing soluble factors since these macrophage-derived factors inhibit the development of CPI and splenomegaly when injected in vivo. Development of this type of suppressor activity appears to be unigenic, recessive, and controlled by genes within the Ig complex as determined by breeding studies using Fl and F2 mice and mice congenic at the Ig locus.

Studies on the nature of the suppression of cytotoxic T lymphocyte (CTL) function exerted by spleen cells from mice infected with Mycobacterium lepraemurium were discussed by W. Bullock (co-workers D. Nickerson and R. Havens from the University of Kentucky College of Medicine). The capacity of splenic lymphocytes from M. lepraemurium-infected C3H/Anf mice to express a CTL response to allogeneic target cells is progressively impaired during the course of the infection that terminates in death at approximately 24 to 26 weeks after intravenous injection of 10^8 M. lepraemurium. Spleen cells from infected mice also suppress the CML response of normal spleen cells in cocultures; they do not induce a temporal shift in the optimum peak of CML activity nor can the suppressor effect be overcome by increasing the degree of alloantigenic stimulation. Evidence was presented demonstrating that the suppressor cells act upon the recognitive and proliferative phases of the CTL response. Furthermore, recognition of target cell antigens by fully differentiated cytotoxic cells is not blocked nor is the killing mechanism. Both nylon-adherent and nylon-passed cells manifest suppressor activity as do G-10 passed cell populations. Suppression by the nylon-adherent cells is not removed consistently by multiple incubations on plastic or by double treatment of the cells with anti-Thy 1.2 and

anti-IgG plus complement. High dose x-irradiation (3000R) of the nylon-
adherent population partially reverses the suppressor activity. Conversely,
the suppressor activity of nylon passed cells can be reversed substantially by
low dosage x-irradiation (500R). These studies suggest that suppressor activity by spleen cells from infected mice may be mediated by two different cell
populations, one composed of T lymphocytes and the other composed of cells with
macrophage-like properties that adhere poorly to plastic surfaces. Confirma-
tory studies are in progress.

Contributing to further discussion of the alterations in immunity induced by
infection, M. Hayes and F. Kierszenbaum of Michigan State University reviewed
their studies on the kinetics of cell responses to polyclonal activators during
the acute and chronic stages of experimental Trypanosoma cruzi infection in CBA/
J and BALB/cJ mice. Mice receiving 25 parasites intraperitoneally survived the
acute stage of infection and entered into a chronic stage. Responses to T-cell
specific mitogens (concanavalin A (con A) and phytohemagglutinin (PHA) and to
endotoxic lipopolysaccharide (LPS), a B-cell mitogen were suppressed significantly as early as day 5 after infection with minimal or insignificant responses observed on days 15 and 20 respectively. Spleen cells of mice in the
chronic stage (sacrificed on days 54 and 90 respectively after infection) exhibited normal responses to all mitogens.

During the acute stage of infection, the absolute number of splenic B cells
was increased greatly but returned to normal levels in the chronic stage. Conversely, the absolute numbers of T cells decreased to very low levels by day 20
of the acute stage and returned to normal levels in the chronic stage. These
observations suggested that immune response may be an important factor in modu-
lating the transition from the acute to the chronic stage in experimental T.
cruzi infection.

Suppression of the responses to con A and PHA during the acute stage of infection could be attributed at least in part to a decrease in the absolute

number of T cells in the spleen; however, the presence of an active suppressive phenomenon could not be excluded. Likewise, suppression of the cellular responses to LPS despite increased numbers of B cells in the spleen might be attributed to the presence of suppressor cells or to an intrinsic defect in B cells.

The final contribution to the workshop was made by J. Mansfield of the University of Louisville School of Medicine who presented data to show that C57BL/6 mice chronically infected with the human pathogen, Trypanosoma rhodesiense, exhibit a marked depression in cytotoxic T lymphocyte (CTL) function in vitro and in vivo. The CTL responses by spleen cells from infected animals are depressed before a significant loss of lymph node CTL function can be measured by in vitro studies.

Of interest is the fact that infected mouse spleen or lymph node cells are incapable of passively transferring immunosuppression to normal mouse CTL cultures. Nevertheless, when spleen or lymph node cells from infected mice are enriched for T cells by nylon wool column passage, CTL function is restored. The restoration of CTL activity by removal of nylon adherent cells can only be achieved during earlier stages of infection since purified T cells from infected animals exhibit intrinsic unresponsiveness after the 25th day of infection.

Dr. Mansfield suggested that the depression of trypanosome-infected mouse CTL function may be due to (a) an early anatomical dilution of cytotoxic T cells by B cells and macrophages (a phenomenon that occurs more rapidly and is more pronounced in the spleen) or (b) acquisition of an intrinsic unresponsiveness in the CTL assay by splenic and lymph node T cells from infected mice that is unrelated to any detectable suppressor cell population that has as its target the cytotoxic T lymphocyte.

LYMPHOCYTE-DERIVED FACTORS

Jerry A. Bash, Ph.D.
Immunologic Oncology Division, Department of Pediatrics,
Georgetown University, School of Medicine, Washington, D.C., U.S.A.

The chairman opened the discussion with an overview of the types of
soluble non-antibody lymphocyte-derived factors which have been described in
the literature and an attempt at classification was made. Two major
categories were suggested: effectors and regulators. The effector group
included primarily those antigen-activated T lymphocyte products which have
been shown to be directly involved in the efferent phase of cell-mediated
immunity - the classic lymphokines. These were further subdivided on the
basis of the target cells affected by these mediators, including macrophages
(e.g. migration inhibitory factor, MIF), lymphocytes (e.g. mitogenic factor,
MF), granulocytes (e.g. leukocyte inhibitory factor, LIF), eosinophils (e.g.
eosinophil chemotactic factor, ECF) and fibroblasts (e.g. lymphotoxin, LT).
The limitations of such a classification scheme were discussed. Certain
factors needed to be considered separately, such as transfer factors which may
be antigen-dependent molecules which must be extracted from human lymphocytes,
and the interferons which may play both effector and regulatory roles. It was
also decided that dialyzable mediators such as cyclic nucleotides, although
ultimate effectors of extracellular stimuli should not be included in the
present classification. The problems of separating functionally heterogenous
supernates into distinct molecules was also briefly discussed using the

example of the inability to clearly separate migration inhibitory factor
(MIF), macrophage activating factor (MAF) and soluble immune response
suppressor (SIRS).

Regulatory factors were subdivided into helper factors and suppressor
factors and further into antigen-specific and nonspecific factors and still
further on the basis of structural or functional relationship to the I region
of the H-2 major histocompatibility region in the mouse. It was pointed out
that these factors could act on a variety of T-cell subsets, directly on
B-cells or indirectly through macrophage presentation.

Discussion of poster presentations began with a brief review by R.L
Fairchild of his work with J.M. Lammert on modulation of macrophage Fc
receptor mobility by a MIF-rich supernatant. Using fluorescein-labelled staph
protein A to study the distribution of cytophilic immunoglobulin on mouse
peritoneal macrophages it was found that cap-formation was significantly
inhibited by MIF-rich supernate. The blocking of supernate activity with
fucose confirmed the phenomenon as MIF-mediated. The authors proposed the
interesting hypothesis that a decreased rate of capping and internalization of
cytophilic antibody might enhance antibody-dependent effector function. The
paradoxical effects of MIF-MAF i.e. decreased membrane mobility but increased
biochemical activity was discussed in terms of localization of activated
macrophages.

Two abstracts from B.W. Papermaster's lab were briefly discussed in the
absence of the authors. The problem of lymphokine binding to serum protein
was discussed in relation to the abstract of McEntire et al. dealing with
-2-macroglobulin and MAF. The abstract of Gilliland et al. dealing with
lymphoblastoid cell line supernates demonstrating tumor regressive properties
was discussed in the context of synergistic effects of multiple lymphokines.

The emerging evidence for interferon in anti-tumor activity was mentioned. This concluded the treatment of effector factors and the discussion turned to regulatory factors.

J. Francisco from J.R. Battisto's lab presented evidence that T-cell Replacing Factor (TRF) which enhances antibody synthesis in vitro selectively suppresses IgG antibody synthesis in vivo. This very interesting finding emphasized the classic dichotomy between in vitro analysis and in vivo relevance. Evidence was presented suggesting the helper T-cell as a target for TRF's suppressive effect in vivo. Differences in cell proliferation kinetics in vitro and in vivo was discussed in an attempt to resolve the paradox.

Mary Parker next presented very interesting findings from G. Sonnenfeld's lab involving the bidirectional regulatory role of interferon which indicated that even within in vitro systems, species differences occur. An apparent reversal in the regulatory kinetics between murine and human systems was reported in which interferon before SRBC sensitization suppressed antibody responses of murine spleen cells while enhancing human peripheral blood lymphocyte responses. Interferon exposure at the same time or after sensitization had the opposite effect in both species. Since interferon regulatory effects were presumed to be species specific as the anti-viral property is, cross-species comparison data were not available. The question of lymphocyte subpopulations peculiar to different tissue was also difficult to approach since human spleen cells and mouse peripheral blood cells were not readily obtainable. A general discussion of the multiple effects of interferon(s) on a variety of target cells followed.

The final presentation by D.A. Vallera from J.R. Schmidke's lab was perhaps the most provocative in that the data supported the concept of a suppressor B-cell capable of inhibiting generation of T-cell-mediated cytotoxicity. Although the B-cell nature of the suppressor cell was supported by its stimulation by LPS, nylon wool adherence, X-ray sensitivity and absence in C3H/HeJ mice, none of these criteria conclusively eliminate macrophage precursors. Support for the functional subset of suppressor B-cells was provided from Dr. Singhal's group which has reported somewhat analogous findings with bone marrow cells and positively identified surface immunoglobulin-bearing cells. A soluble B-cell product (presumably non-immunoglobulin) was reported to mediate this suppression, but characterization was incomplete.

The session was concluded by a general discussion of the increasingly heterogenous array of lymphocyte-activation products and the continued need for comparison, chemical characterization and establishment of in vivo relevance.

AUTOIMMUNITY, IMMUNODEFICIENCY AND IMMUNOLOGY OF AGEING

THOMAS G. WEGMANN[*] and ROY SUNDICK[**]
*Department of Immunology, University of Alberta, Edmonton, Alberta,
 Canada; **Department of Immunology and Microbiology, Wayne State University,
 Detroit, Michigan, U.S.A.

The unifying theme of this workshop was the immunoregulation involved in the processes of autoimmunity and immunodeficiency as a function of the age of the animal. A number of studies have implicated a loss of T suppressor cell activity in autoimmunity, the classical case being systemic lupus erythematosis and the mouse counterpart, the NZB/NZW mouse model. Other studies have indicated that T suppressor cells may also be involved in immunodeficient states. For example, one can find suppressor cells capable of preventing immunoglobulin synthesis in common variable hypogammaglobulinemia as well as in bursectomized chickens. Given this recent emphasis on regulatory mechanisms in reference to autoimmunity and immunodeficiency, most of the papers presented at this session dealt with various aspects of immunoregulation. The papers themselves fall into two broad categories. The first category involves autoimmunity. There were four papers relating to systemic autoimmunity presented in the first part of this session, three dealing with mouse models and one dealing with a human model of global autoimmunity. The rest of the papers under the autoimmunity section involved specific organ systems. There were three papers involving autoimmunity in the nervous system, three papers concerned with thyroid auto-immunity and two papers examining autoimmunity in the reproductive tract. In the second borad category, that of immunodeficiency, three papers were presented.

1) Autoimmunity

   A) Systemic autoimmunity. The first paper was presented by Dr. T. Hsiaso of the Cleveland State University and Dr. R.S. Krakauer of Cleveland Clinic Foundation, and involved a therapy model for the NZB/NZW F_1 hybrid autoimmune model. Twenty-week old animals were first treated with prednisolone to induce a remission of their lupus-like disease, and then treated with Con A stimulated spleen cell supernatants in an attempt to induce immunosuppression. This treatment reduced the proteinuria and the severity of the immune complex glomerulone-phritis of these animals, but without affecting the anti-DNA antibody levels. Since others have shown that reduction of anti-DNA antibodies leads to a reversal of the glomerulonephritis seen in this disease, and elution of anti-bodies from NZB kidneys yields predominantly anti-DNA antibodies, the present results will require a reexamination of the pathogenesis of the glomerulone-phritis seen in this model system.

The paper by Drs. E.H. Greeley, and M. & D. Segre represented an attempt to re-late a non T suppressor cell which appears in regenerating spleens following large doses of cyclophosphamide to self tolerance. The conjecture was that if this suppressor cell were involved in the maintenance of self tolerance, it might be missing in older autoimmune NZB mice. However, the activity of the cell actually increased in older NZB mice as well as in older mice of other strains. The appearance of this cyclophosphamide-induced suppressor cell is reminiscent of suppressor cells which appear after treating mice with Stron-tium[89]. One supposes that these agents create a cellular "vacuum" which can then be filled by minority cell populations.

Drs. D. Wilson, and H. Braley-Mullen discussed the MRL/1 and MRL/N mice, both of which developed glomerulonephritis. In the former case it develops earlier and is more severe. The conjecture is that helper cell function is de-ficient in the MRL/1 mice. In order to evaluate this idea, both young and old mice of both strains were challenged with a thymus dependent antigen (horse red blood cells) and a thymus independent antigen (Type III pneumococcal polysaccha-ride). All responses were normal, except for those of the older MRL/1 mice, which showed a defective response to the thymus dependent antigen. The cellu-lar basis for this defect is not yet known, but will be the subject of future investigation.

Systemic autoimmunity in humans was investigated by J. Sundeen, T. Alexander, A Scherbel, and R. Krakauer of the Cleveland Clinic Foundation. They studied *in vitro* responsiveness to pokeweed mitogen (PWM) in patients with progressive systemic sclerosis (PSS), in which various auto-antibodies as well as abnormal T cell function have been demonstrated. They found that PSS B-cells, in the presence of normal T-cells, synthesized less IgM in comparison to controls. Also PSS T-cells were more able to help normal B-cells in synthesizing IgM in response to PWM than normal T-cells. The authors concluded that the PSS pa-tients have abnormally low B-cell function along with increased T helper func-tion. One caution raised in the discussion on this work is the possibility that allogeneic effects, as characterized in the mouse, may be at work here, since there was no attempt to control for HLA type; thus one could be dealing with abnormal induction of IgM synthesis.

B) Nervous system autoimmunity. Drs. C. MacPherson of the University of
Western Ontario and I. Ramshaw of the University of Saskatoon described a system in which bovine spinal cord protein (BSCP) can prevent the encephalitogenic
response of guinea pigs to bovine myelin basic protein (BMBP). By using a
lymph node proliferation assay, they showed that there is immunological crossreaction between BSCP and BMBP and they conjectured that this crossreaction may
eventually explain the anti-encephalitogenic properties of the BSCP. Drs. H.
Rauch, M. Katar and I. Montgomery of Wayne State University described the isolation of two MBP components from mouse brain. One is designated "large", containing between 160 to 164 immuno acid residues while the other is called
"small", containing 114 to 115 residues. These proteins both show a migration
pattern similar to rat MBP's, and only the large protein induces allergic
encephalomyelitis in guinea pigs and mice.

Drs. J. Holda, A. Welch and R. Swanborg of Wayne State University indicated
that experimental allergic enchphalomyelitis (EAE) can be induced in Lewis rats
by incubating spleen cells *in vitro* with basic protein, followed by adoptive
transfer of these cells into syngeneic rats. Furthermore, cells from rats challenged for 12 to 33 days *in vivo* can also induce disease upon adoptive transfer
to syngeneic recipients. Interestingly enough, the adoptive transfer situation
differs from active immunization in that the latter leads to resistance upon
subsequent challenge whereas the former leads to heightened disease upon secondary challenge. This suggests that suppressor cells do not survive adoptive
transfer, and can potentially lead, *via* mixing experiments, to characterization
of the suppressor cells.

C. Thyroid Autoimmunity. Drs. H. Lillehoj and N. Rose of Wayne State
University indicated that the susceptibility to experimental autoimmune throiditis in rats is under complex genetic control. The data suggest that sex chromosomes may be involved in susceptibility, but the numbers had not yet reached
significant levels. MHC genes do not seem to be involved in the control
mechanism(s). Drs. I. Okayasu, Y. Kong and N. Rose presented evidence that
rules out the effects of sex hormones on experimental autoimmune thyroiditis in
mice. They did this by castrating both male and female mice and showing that
this procedure had no influence on the development of the disease.

Auto-immune thyroiditis in obese strain chickens was the model studied by
Drs. A. Sanker and R. Sundick of Wayne State University. These chickens spontaneously develop autoimmune throiditis, beginning at three weeks of age. One
can significantly delay the disease by a single injection of 1 mg of thyroglobu-
lin at hatching. Subsequent injections give no more prolongation than the

single injection. Since the chicken is easily amenable to *in ovo* injections it will be most interesting to see what effect earlier injections have on this disease process.

D) Reproductive Tract Autoimmunity. Drs. Z. Marcus and E. Hess of the University of Cincinnati described a system in which human seminal plasma or spermatozoa added to human lymphocyte cultures can inhibit both base line as well as PHA induced [3]HT incorporation and E-rosette formation. This observation and others like it may provide an explanation as to why females in general do not react immunologically against sperm.

Drs. H. Lipscomb, J. Breitkreutz and J. Sharp of University of Nebraska presented data which indicated that neonatal thymectomy of Lewis rats leads to a high incidence of spontaneous orchitis as manifested by decreased testes weights, increased sperm agglutinating antibody, and testicular histopathology. A less severe disease can be produced in naive intact recipients by transferring spleen cells from affected donors. It will be of interest to attempt the adoptive transfer of normal T cells to thymectomized donors to further characterize the normal immune regulation involved in preventing orchitis.

2) Immunodeficiency and Ageing

A method for the detection of heterozygotes for congenital X-Linked agammaglobulinemia was presented by Drs. L. Thompson, G. Boss, H. Spiegelberg and J. Seegmiller of the University of California at San Diego and the Scripps Clinic. Patients with congenital X-Linked agammaglobulinemia have reduced ecto-5'-nucleotidase activity in their peripheral blood lymphocytes. These workers were unable to identify female carriers by assaying this enzyme activity in either unfractionated peripheral blood lymphocytes or in T or B cell enriched lymphocyte subpopulations. Surprisingly, lymphoblastoid cell lines established by Epstein Barr virus transformation from obligate heterozygotes showed markedly less enzyme activity than controls (less than 10% in the studies reported): These cell lines were also deficient in surface immunoglobulin, and the authors suggested that they may represent transformation of immature B cells. This fascinating observation provoked a good deal of discussion. One principle question is why the heterozygous carriers showed such low enzyme activity rather than 50% of normal as might be expected on the basis of random X chromosome activation. The possibility emerged that the affected B cells in the heterozygote are arrested at an immature stage and these cells are more susceptible to transformation by Epstein Barr virus than cells with the normal X chromosome in active form, although that point is by no means proven at this time.

Dr. R. Cross of the University of Kentucky presented evidence that rats with anterior hypothalamic lesions show changes in lymphoid cellularity and a decreased responsiveness to Con A, with the maximal effect at four days after lesioning. He was able to rule out corticosteroid release as an explanation. This represents another attempt to examine the poorly understood neuroendocrine-immune system axis, which will undoubtedly be the subject of much future investigation.

The final paper, presented by Drs. K. Rao and S. Schwartz from the University of Michigan involved an addition to their previous observation that zinc differentially affects lymphocytes from older individuals with respect to lectin stimulation. These workers reported indirect evidence that what might be involved is aging of the microfilament structures of these lymphocytes, because zinc also affected the cytochalasin B dose response curve of lymphocytes stimulated by Con A. Since cytochalasin B is thought to interfere with the functioning of microfilaments, this indirectly implicates these structures in the ageing process. Drs. Rao and Schwartz intend to study microfilaments in a more direct fashion to attempt to implicate them in the ageing process.

Altogether the papers presented in this workshop represented a spectral sampling of the many different approaches being taken to understand the regulation involved in autoimmunity and immunodeficiency as a function of age. While it is clear that complex immunoregulatory processes are disturbed in both disease states, the exact role of deranged immunoregulation in the etiology of these disorders remains to be clarified.

CELL-CELL INTERACTIONS AND THE ROLE OF THE MACROPHAGE

IN THE IMMUNE RESPONSE

JOHN E. NIEDERHUBER[+] AND STEVE LERMAN[++]
[+]Department of Microbiology/Immunology and Department of Surgery,
University of Michigan Medical Center, Ann Arbor, Michigan, U.S.A.,
[++]Department of Immunology and Microbiology, Wayne State University,
Detroit, Michigan, U.S.A.

INTRODUCTION TO WORKSHOP

Antigenic stimulation of immunocompetent lymphoid cells to effect an immune response requires interactions among T-lymphocytes, B-lymphocytes and antigen-presenting macrophages.[1-3] It is clear that gene products of the major histocompatibility complex (MHC) have a critical role in these cell interactions.[4-13] Evidence for the involvement of MHC products was initially provided by the experiments of Katz and co-workers demonstrating the requirement for MHC identity between antigen-primed helper T-cells and B-cells.[6] Similarly, Rosenthal and Shevach showed that proliferating T-cells required the presence of syngeneic macrophages.[7] Interpretation of these early experiments was complicated by the allogeneic effects of mixing histoincompatible cells.

Evidence that interaction of the involved cell populations was more than a simple requirement for MHC identity came from experiments by Pierce and colleagues. They observed that GAT primed T-cells only responded in-vitro to macrophages bearing the same H-2 products as macrophages used to prime the T-cells.[9] Similarly, Erb et al. found that F_1 T-cells activated to antigen in-vitro in the presence of parental strain$_A$ macrophages could not activate B-cells of the other parent (strain$_B$), whereas F_1 B-cells gave good responses.[8,12] The inability of the B-cells from parent$_B$ to respond was overcome by adding macrophages from parental strain$_A$.

In order to eliminate potential allogeneic effects, a number of investigators have asked questions concerning restriction of cell interaction using T-cells derived from radiation induced chimeras.[14-23] Such experiments demonstrated that strain$_A$ T-cells which had differentiated in an (A x B)F_1 chimeric environment could support the response of either strain$_A$ or strain$_B$ B-cells. Thus, it began to emerge that the restrictions observed were imposed by the thymic environment on the T-helper cell during T-cell differentiation and were not the result of antigen priming.

Experiments utilizing antigen stimulation under Ir gene control and chimeric mice constructed of responder and nonresponder strains have been more difficult to interpret.[15,18,23] In these experiments, either high responder or low responder helper T-cells obtained from an (HR x LR)F_1 chimera could provide T-help in collaboration with high responder macrophages and B-cells. However, (HR x LR)F_1 T-helper cells obtained from a low responder chimera could not provide help.

A different approach to the questions of macrophage T-cell interactions has involved the use of alloantisera produced against I-region determinants. Anti-Ia serum has effectively blocked the antigen induced proliferative response of primed T-cells in both the guinea pig[24] and the mouse.[25] Recently, the antigen specific T-cell proliferative response to GLØ, which is under dual Ir gene control, was shown to be blocked by anti-Ia serum specific for either I-A or I-C subregions. These results suggested that Ia antigens were the phenotypic expression of Ir genes.

Additional evidence for the role of I region products in macrophage-T cell interaction, comes from experiments using macrophages to select for antigen-specific T-cells. These experiments demonstrated that treating the macrophage population with anti-Ia serum prevent the macrophages from isolating antigen specific T_H-cells required for the secondary response to TNP-KLH.[27]

Anti-Ia serum has also been effective in blocking macrophage function in the primary in-vitro antibody response to erythrocyte antigens. As in the antigen-specific T-cell proliferative responses, an Ia^+ subpopulation of macrophages was required.[28,29] The results of these experiments demonstrated that only antibodies specific for the I-J subregion could effectively block macrophage recognition by T_H-cells.[30] In contrast to the antigen induced T-cell proliferative response, antibodies specific for the I-A subregion could not block macrophage-T cell interaction in the primary antibody response. The anti-Ia antibodies reactive with determinants of I-A, I-E, or I-C subregions always required complement and therefore deletion of the required macrophage subpopulation in order to inhibit the primary in-vitro antibody response.

The specific nature of the anti-I-J serum blocking of

macrophage-T cell interaction was further confirmed when these antibodies were used to pretreat F_1 macrophages.[13] Effective blocking of the response was only observed when the anti-I-J antibodies were specific for the phenotype of the T_H-cells used to reconstitute the response. Anti-I-J serum blocked F_1 macrophages were perfectly capable of interacting with F_1 T-cells or T-cells of the parental phenotype which was not blocked by the anti-I-J serum. These experiments were interpreted as demonstrating the requirement for genetic restriction at the T-cell level for syngeneic I-J region determinant(s) in the primary in-vitro antibody response.

The different results obtained with antigen specific T-cell proliferation versus primary in-vitro antibody responses are not incompatible when one considers the non-specific nature of T-cell proliferation and the level of response - primary versus secondary. The model in which anti-I-J region antibodies block macrophage-T cell interaction during the initial step of antigen induction of the antibody response, requires steps of cell differentiation at the T-cell level, cell interaction, cell triggering at the B-cell level, B-cell differentiation and clonal expansion. The end result of which is a highly specific clone or clones of B-cells producing a population of antibodies directed at determinants on the stimulating antigen. Experimentally there is no indication that T-cell proliferation in response to antigen can be equated with the induction of T-cell help.

A final word of caution is necessary to conclude the introduction to this workshop. The interpretation of experiments designed to evaluate mechanisms of cell interaction and genetic restriction depend heavily on the ability to prepare and use pure populations of antigen-presenting macrophages, T-helper cells and antibody-forming cells. Such purifications are always difficult and minor contamination of unsuspected functionally different cells may hinder correct interpretation of results. The T-cell population clearly functions in a regulatory capacity, thus experiments must take into account the possibility of unrecognized selection or enrichment for one T-cell subset over the other. Despite these concerns much has been learned about the interaction of macrophages and T-cells.

WORKSHOP SUMMARY

Dr. F. R. Cochran presented data indicating that carrageenan-induced suppressor macrophages secreted an inhibitor of T-cell proliferation. The work was conducted in collaboration with Drs. J. Vago and J. A. Bash in the Department of Environmental Health and Microbiology at the University of Cincinnati. They found that supernates obtained from purified rat peritoneal exudate macrophages after 72 hours culture with 1-10 μg of carrageenan profoundly inhibited splenic T-cell proliferation in response to phytohemagglutinin (PHA). Decreased inhibitor production in the presence of indomethacin was interpreted as suggesting prostaglandin mediated suppression. Rats receiving carrageenan orally showed similarly depressed spleen cell responses to PHA which were dependent upon adherent cells. Supernates from peritoneal macrophages of carrageenan-fed rats were highly suppressive to spleen cell PHA responsiveness, suggesting a common mechanism of macrophage mediated suppression operating _in vitro_ and _in vivo_.

Dr. P. S. Duffey of the Department of Microbiology, the University of Texas Health Sciences Center at San Antonio, addressed the problem of the roles for 2-mercaptoethanol (2 ME), serum factors and macrophages in the responses of lymphocytes to B- and T-cell mitogens. He reported that the addition of 2 ME to serum-containing C58 mouse splenocyte cultures enhanced the response to B- and T- cell mitogens as measured by tritiated thymidine ([3]HtdR) incorporation. 2 ME alone did not enhance the mitogen response, nor did 2 ME itself act as a mitogen in this system. Some lots of serum contained relatively high amounts of a 2 ME-activated factor, enhanced [3]HtdR incorporation, while others contained a 2 ME-activated inhibitor of HtdR incorporation. Addition of silica to cultures containing 2 ME and serum abrogated [3]HtdR incorporation. This inhibition was blocked by adding polyvinylpyridine N-oxide (PVNPO) to the cultures or was reversed by replacing silica-killed splenic macrophages with PVNO-protected adherent peritoneal exudate cells. These findings were interpreted by Dr. Duffey as indicating that 2 ME does not replace the function of macrophages and further, that macrophages are required for both B- and T- cell mitogen-induced incorporation of [3]HtdR by C58 splenic lymphocytes.

A presentation was made by Dr. E. M. Allen of research performed in collaboration with Drs. D. J. Schrier, J. Sternick and V. L. Moore of the Research Service, Wood Veterans Administration Medical Center and the Department of Medicine and Pathology, Medical College of Wisconsin. Earlier work from this group had demonstrated that certain strains of mice developed chronic pulmonary inflammation (CPI) and splenomegally after IV injection with killed BCG. It had also been shown that spleen cells from BCG-injected-C57BL/6 (BCG-B6) responder mice were markedly suppressed in their responses to PHA and LPS, and that Thy-1 negative-adherent spleen cells from these animals were suppressive for normal B6 spleen cells. Furthermore, it was found that BCG-B6 mice did not produce a potent primary antibody response to sheep erythrocytes (SE) and were unable to generate significant delayed hypersensitivity (DH) to SE. They now reported that spleen cells from BCG-B6 mice elaborated materials in culture which: 1) suppressed PHA responses of normal syngeneic spleen cells, 2) reduced DH to SE, and 3) suppressed the development of CPI. The spleen cells which produced these factors were Thy-1 negative and adhered to plastic petri plates.

Dr. Thomas Huff discussed the development of a primary culture technique for isolating trypsin resistant macrophage-like cells from the skin of 17-19 day old murine embryos. The research was conducted in collaboration with Drs. J. Snellhaas, G. Stelzer, D. Justus and J. Wallace in the Department of Microbiology and Immunology, University of Louisville School of Medicine. The cells which they obtained stained positively for nonspecific esterase and exhibited surface receptors for Fc-IgG and complement components. In addition they stained positively for membrane ATPase activity, and a large proportion expressed Ia antigens. The above properties were consistent with those already described for Langerhan's cells (LC). Although the cells described by Dr. Huff were strongly phagocytic, LC are reported to be less capable of phagocytosis. However, Dr. Huff examined phagocytic capabilities in vitro whereas earlier studies were based on in vivo or in situ techniques. In addition, whereas LC are characterized by unique trilaminar rod-shaped granules, such granules were absent in the cells isolated by Dr. Huff. However, since cells bearing these granules have not been reported to be present in the skin of mice younger than 6 days of age, it was pointed out by Dr. Huff that it was not unexpected when such granules were not found in cultured embryonic skin. Dr. Huff's technique, therefore, seems to be useful in isolating Langerhan-like cells from embryonic skin.

Dr. T. J. Gorzynski of the Department of Microbiology of the State University of New York at Buffalo, in association with Dr. M. B. Zaleski, presented data examining the humoral response to Thy-1 alloantigens and interpreted the resulting data as a possible case for associative recognition. Mice were immunized with Thy-1 disparate thymocytes 6 days prior to analysis of their spleen cells for plaque forming cells specific for the immunizing thymocytes. Good responsiveness required compatibility between donor and responder at the K D, and L regions of the H-2 complex while compatibility at the I, S, and G regions was unnecessary. Responses of H-2 heterozygous mice were lower than those of the corresponding H-2 homozygous parental strain when immunized with thymocytes from H-2 homozygous mice compatible with this parent. Responses of H-2 heterozygous F_1 hybrids were often significantly different when immunized with thymocytes from donors compatible with either one or the other parent. The lower and often asymmetric responses of F_1 hybrids were interpreted as being consistent with the concept of their possessing randomly generated clones capable of recognizing Thy-1 antigen in the context of H-2 molecules from either parent. Although Dr. Gorzynski felt that these data were an example of associative recognition and that antigenic competition could be ruled out, several discussants felt this point was not totally clear.

Dr. P. Kavathos and her associates, Drs. F. H. Bach and R. DeMars of the Immunobiology Research Center and the Laboratory of Genetics of the University of Wisconsin, described a technique for obtaining variants of human lymphoblastoid cell lines that have lost expression of all HLA antigens coded for by a single haplotype. This was achieved by using ionizing irradiation (300 R) as a mutagenic agent. Anti-HLA-B8 serum and rabbit complement were used to select for variants of a human lymphoblastoid cell line that carried HLA haplotypes HLA-A1, B8, DRW3 and HLA-A2, B5, DRW1 and that were heterozygous at the glyoxalase I locus. Selection was imposed 5 days post-irradiation. Variants were obtained with an incidence of 4.1×10^{-5} compared to a spontaneous incidence of 0.5×10^{-5}. Variants that no longer expressed HLA-B8 never lost expression of trans alleles, but 13 of 18 variants had lost expression of additional Cis linked HLA- and GL∅ alleles. This approach could prove useful in consistently obtaining phenotypically homozygous cell lines from heterozygous lines which could then be used for the same purposes as are the more difficult to obtain conventional HLA homozygous lymphoblastoid cell lines.

REFERENCES

1. Mosier, P.E. (1967) Science, 158, 1573.
2. Lipsky, P.E. and Rosenthal, A.S. (1975) J. Exp. Med. 131, 138.
3. Lipscomb, M.F., Ben-Sasson, S.Z., and Unr, J.W. (1977) J. Immunol. 118, 1748.
4. Green, I., Paul, W.E., and Benacerraf, B. (1966) J. Exp. Med. 123, 859.
5. McDeVitt, H.O. (1968) J. Immunol. 100, 485.
6. Katz, D.H., Hamaoka, T., and Benacerraf, B. (1973) J. Exp. Med. 173, 1405.
7. Rosenthal, A.S. and Snevach, E.M. (1973) J. Exp. Med. 138, 1194.
8. Erb, P. and Feldman, M. (1975) J. Exp. Med. 142, 460.
9. Pierce, C.W., Kapp, J.A., and Benacerraf, B. (1976) J. Exp. Med. 144, 371.
10. Kappler, J.W. and Marrack, P.C. (1976) Nature (Lond.) 262, 797.
11. Swierkosz, J.E., Rock, K., Marrack, P., and Kappler, J.W. (1978) J. Exp. Med. 147, 554.
12. Erb, P., Meier, B., Kraus, D., von Boehmer, H., and Geldman, M. (1978) Eur. J. Immunol. 8,786.
13. Niederhuber, J.E. and Allen, P. (1980) J. Exp. Med. (In Press).
14. von Boehmer, H., Hudson, L., and Sprent, J. (1975) J. Exp. Med. 142, 989.
15. Kappler, J.W. and Marrack, P. (1978) J. Exp. Med. 148, 1510.
16. Sprent, J. (1978) J. Exp. Med. 147, 1838.
17. Hodes, R.J., Hathcock, K.S., and Singer, A. (1980) J. Immunol. 124, 134.
18. Marrack, P. and Kappler, J.W. (1979) J. Exp. Med. 149, 780.
19. Kindred, B. (1975) Cell. Immunol. 20, 241.
20. Katz, D.H., Skidmore, B.J., Katz, L.R., and Benacerraf, C.A. (1978) J. Exp. Med. 148, 727.
21. Sprent, J. and Bruce, J. (1979) J. Exp. Med. 150, 715.
22. Singer, A., Hathcock, K.S., and Hodes, R.J. (1979) J. Exp. Med. 149, 1208.
23. Hedrick, S.M. and Watson, J. (1979) J. Exp. Med. 150, 646.
24. Snevach, E.M., Green, I., and Paul, W.E. (1974) J. Exp. Med. 139, 679.
25. Schwartz, R.H., David, C.S., Sach, D.H., and Paul, W.E. (1976) J. Immunol. 117, 531.
26. Schwartz, R.H., David, C.S., Dorf, M.E., Benacerraf, B., and Paul, W.E. (1978) Proc. Natl. Acad. Sci. USA, 75, 2387.
27. Swierkosz, J.E., Marrack, P., and Kappler, J.W. (1979) J. Immunol. 123, 654.
28. Niederhuber, J.E. (1978) Immunological Rev. 40, 28.
29. Niederhuber, J.E., Allen, P., and Mayo, L. (1979) J. Immunol. 122, 1342.
30. Niederhuber, J.E. and Allen, P. (1980) J. Immunol. (In Press).

IMMUNOGLOBULIN GENETICS, STRUCTURE AND FUNCTION

W. CAREY HANLY
Department of Preventive Medicine and Community Health, University of Illinois
at the Medical Center, Chicago, Illinois 60680

Since the early 1950's when antibody activity was first associated with the
gamma globulin fraction of serum, immunologists have continuously sought greater understanding of these enigmatic molecules that are so central to host
defense. Investigations of immunoglobulins have proceeded along many lines
during the intervening years, and it is through contributions from diverse
fields including biochemistry, genetics, molecular and cellular biology that
we have reached our present state of knowledge. This workshop entitled
"Immunoglobulin Genetics, Structure and Function" brought together a group of
investigators with seemingly diverse specialties; yet each came because of a
central interest in the antibody molecule. The presentations and discussion
of the preceding symposium, "Immunoglobulin Genes," and the variety of topics
discussed in this workshop emphasize the continuing need for many strategies
of study of immunoglobulin molecules.

Antibody diversity and the mechanism for generation of diversity have been
a major focus of study for many years. Combinatorial joining of heavy (H) and
light (L) chains and of gene segments for each of these chains now appears to
account for much of the diversity which has been observed or proposed. However,
it has also been clear for a number of years that some combinations of H and L
chains are preferred over others and that in all probability not all combinations are useful. Fred Stevens, discussing work done in collaboration with
Florence Westholm and Marianne Schiffer from Argonne National Laboratory and
with Alan Solomon from the University of Tennessee Center for the Health
Sciences, suggested a basis for preferential combination of immunoglobulin (Ig)
chains. Their studies of the self-association properties of κ_I Bence-Jones

proteins showed that differences in dimerization constants ranging from less than 10^3 M^{-1} to greater than 10^6 M^{-1} could be related to amino acid sequence; and that the major relevent residue was at position 96, a position known to be an interchain contact residue and to be encoded by the N-terminal portion of the J-segment of the genome. A hydrophobic and/or aromatic residue at position 96 was related to a high dimerization constant, whereas a polar amino acid at position 96 was related to a low dimerization constant. Presumably this type of correlation extends to combinations of H and L chains. It was suggested that the "J genes" which contribute portions of the genetic information for the third hypervariable regions of Ig polypeptide chains have evolved separately from the major V-region segments, and in a manner which helps to maximize antibody diversity. Given such information, new questions arise as to the basis of selection of particular J-segments for H and L chains within a single cell.

Further discussion on diversity of antibody molecules and limitations to diversity followed presentations of two studies where genetic factors were given some consideration for basis of restriction of the antibody response. Kathy Grzyb from Bernard Friedenson's laboratory at the University of Illinois at the Medical Center showed some of their results from a comparative study of antibody responses to the hapten p-azobenzoate by outbred rabbits and by members of a line of inbreeding rabbits from the colony of Carl Cohen and Robert Tissot. From isoelectric focusing patterns the specific antibodies or their L-chains appeared to be of restricted heterogeneity, indicating that the number of clonotypes expressed was small. This restricted response appeared to be independent of the coefficient of inbreeding of the recipient and independent of which protein carrier (keyhole limpet hemocyanin or bovine gamma globulin) was used for the immunogen. Other evidence for a restricted antibody response was presented by Marion Fultz from the University of Michigan. Her data suggested that the naturally occurring anti-B (blood group) antibodies among identical individuals (monozygotic twins or triplets) may share some idiotypic determinants. Such data suggest that the idiotype reflects a germ-line gene.

An example of antibody-mediated conformational stablization of antigen was discussed by Mi-Kyung Dong who has worked in conjunction with B.K. Choe and Noel Rose at Wayne State University Medical School to study human prostatic acid phosphatase. Her results suggested that antibody elicited against the whole enzyme could restore catalytic activity to dissociated subunits of the enzyme or to a CNBr-derived peptide. Although previous examples of allosteric effects on antigen by antibody binding have been demonstrated, additional studies should increase understanding of this process and perhaps reveal situations where this process functions *in vivo*. In addition, there could be a broader application of this type reaction to identify active sites of enzymes.

During the rest of the workshop, discussion was directed towards effector functions of antibody molecules. Betty Anne Johnson and Louis G. Hoffmann of the University of Iowa talked about the relationship of the state of disulfide bonding in rabbit IgG antibody and its ability to activate complement. Samples of rabbit IgG anti-DNP which had been reduced under different conditions (aerobic and anaerobic) were assessed for quaternary structure by SDS-polyacrylamide gel electrophoresis and for ability to bind complement and hemolyze TNP-erythrocytes. IgG antibody which had been reduced aerobically, such that no intact inter-heavy chain disulfide bonds were found, retained approximately 50% of its original hemolytic activity. With anaerobic reduction, hemolytic activity was lost even though a significant percentage of the antibody molecules retained intact inter-heavy chain bonds. Combined results suggested that an intra-H chain disulfide bond rather than the inter-H chain bond is critical for maintaining appropriate antibody conformation to effect complement activation. Rabbit IgG has an unusual inter-domain disulfide, from the

$C_\gamma 1$ domain to the hinge, which is a likely candidate for the critical bond. Further studies such as this are needed to increase understanding of antigen-dependent effector functions of antibody molecules. The activation of complement by antibody to effect hemolysis must require a change of conformation of the antibody, but how this is brought about is not clear. Further discussion of possible mechanisms for antigen dependent Fc functions followed Margaret Watanabe's presentation of studies on hemolytic efficiency of rabbit IgM anti-arsanilazo (AA) antibody. This study, done with Joseph Ingraham of Indiana University School of Medicine and B.H. Petersen of Eli Lilly and Company, illustrated a way in which the state of the antigen could influence hemolytic efficiency of antibody. When IgM anti-hapten antibody was tested with sheep cells (SRBC) to which the hapten AA was coupled directly, the hemolytic efficiency was only 25% to 50% the efficiency found when the antibody was tested with SRBC to which the AA hapten was coupled via a tyrosyl·glycyl· glycyl extender arm. The difference seemed to be attributable to the extender arm since fragility indices of the AA-SRBC and the AA-YGG-SRBC were similar when tested with anti-SRBC plus complement. Although the exact way in which the extender arm promoted hemolytic efficiency was not clear, these results promoted a discussion on theoretical aspects of mechanisms of antigen-dependent effector functions as well as on practical aspects of assays involving hapten coupled to erythrocytes. It is clear that further studies relating structure and effector functions of Ig molecules are much needed. Additional data on three-dimensional structures of Fc portions of Ig molecules should provide some basis for more specific identification of those sites involved in interaction with Fc receptors and with complement.

B CELL ACTIVATION AND IMMUNOLOGICAL MEMORY

THOMAS L. FELDBUSH
University of Iowa and The Veterans Administration Medical Center, Iowa City,
Iowa 52242

Two major themes were discussed in this workshop; 1) development and
activation of B cells, and 2) the generation and maintenance of immunologic
memory. The workshop was opened by a brief but eloquent review of the fetal
development of B cells presented by Professor Dennis Osmond. In this review
Osmond presented their work showing that one of the immature B cell stages is
characterized by a cytoplasmic mu positive blast-like cell. Riley and Kuehl
elaborated on this B cell population. They described transformed cells which
are arrested at different stages of B cell development and proposed that this
would provide an ideal model system for the study of the regulation of
synthesis and subcellular localization of Ig during development. Some of these
continuous cell lines (all Abelson murine leukemia virus transformed bone
marrow cells) have the characteristics of pre-B cells in that they are cytoplasmic mu chain positive, Thy 1 negative, Ia negative, complement receptor
negative and sIg negative.

Riley and Kuehle also reported that somatic cell hybrids were made between
the pre-B cell line and plasma cell line in order to determine how Ig synthesis
was affected in hybrids derived from parental cells in different but closely
related stages of B cell development. One hybrid was made between the pre-B
cell and a myeloma which synthesizes and secretes IgG_{2a}. This hybrid expressed
both mu and gamma 2a H chains with each class being assembled to a single light
chain. Another hybrid was made between the pre-B cell and a myeloma which
synthesizes only the gamma 2a H chain, no light chain. Interestingly, many of
the hybrids produced a light chain. The source of the genetic information for

this light chain has not yet been identified.

Miller presented some interesting observations on the blocking of antigen induced B cell tolerance in both neonatal and adult lymphocytes. When T lymphoblastoid cell lines are cultured, they shed Thy-1 gangliosides in vesicles. Miller has found that these gangliosides can interact with neonatal spleen cells and prevent tolerance induction to DNP. Under normal circumstances the predominant B cell population in the neonatal spleen is the sIgM predominant immature cell which Vitetta and co-workers and Klinman and co-workers have shown to be very susceptible to antigen induced tolerance. In a similar manner the gangliosides can prevent high dose antigen tolerance of adult B cells. Miller suggests that under normal circumstances of immunization, the B cell population can interact directly with antigen and thus be rendered tolerant before adequate T cell help is generated. If the T cell gangliosides are shed in high enough concentrations in vivo then they could prevent this premature B cell tolerance. Thus Miller would propose that this is a normal regulatory mechanism allowing for the induction of immune responses rather than tolerance.

Miller also presented data showing that the gangliosides were also active on adult bone marrow cells and memory B cells. The bone marrow cells were fractionated by FACS into mu positive and negative populations. The mu positive cells, like the adult spleen cells could only be rendered tolerant with high antigen dose. The mu negative cell, like the neonatal population was quite susceptible to tolerance induction. Thy-1 gangliosides blocked tolerance in both populations. Likewise, the gangliosides blocked high dose antigen tolerance of memory B cells. This data suggests that the gangliosides may have a very wide range of effects on many lymphocyte subpopulations.

It has been reported by Tittle and Rittenberg and confirmed in our laboratory that thymus dependent (TD) antigen priming stimulates the generation of two separate IgG secreting, memory cell subpopulations. One population can be triggered by a TD form of the antigen and the other population by a thymus independent (TI) form of antigen. Lite and Mullen reported in the workshop that the same compartmentalization may not exist when priming is made with a TI antigen. Using PVP and adoptive transfer, low dose priming resulted in the generation of IgG secreting memory cells which could be triggered with the TD antigen PVP-HRBC. No IgG memory response was seen when PVP alone was used for challenge but the PVP could block the response to PVP-HRBC. Thus, while IgG secreting memory cells responsive to TI antigen can be induced by TD priming, the TI antigen itself is not capable of inducing the formation of this cell population. Furthermore, those memory cells which are induced by TI immunization cannot be triggered by the homologous TI antigen. They will react with the antigen as evidenced by the blocking experiments.

As pointed out by Mullen, caution must be exercised in interpreting these results. For example, the system used in studying this phenomenon may be very important. Mullen has shown that animals primed with the TD form of SIII will respond with an IgG response to the TI form of SIII but only when the intact donor is challenged. In adoptive transfer experiments no IgG is seen. Furthermore, not all TI antigens are the same. In Mullen's experience only PVP (a type II TI antigen) in suboptimal doses induce IgG memory cell development and these memory cells will not respond to the TD antigen in the absence of T cells. No other TI antigen will induce IgG memory. On the other hand, the memory cells induced by DNP-BGG will respond to TNP-B. abortus (Type I TI) without T cell help as shown by Tittle and Rittenberg and in our own experiments.

At present it is probably too early to try to reconcile all these different observations, however, as described below, it is becoming apparent that many different subpopulations of memory cells exist. It is conceivable that the nature of the antigenic trigger (with or without T cell help, strength of the mitogenic signal, epitope density, etc.) may favor or preclude a particular line of differentiation resulting in memory cell populations with a variety of characteristics and triggering thresholds. It was concluded that this area would appear to be a fertile ground for investigation.

The results presented in the workshop by Hobbs, Brooks and Feldbush were assembled, together with the published results of Strober et al., into a model for the generation of memory cells to TD antigens. Early after immunization large sIgM+, CR-, memory cells appear in the antigen draining lymph nodes. Some of these cells remain localized in the nodes while other circulate to distal nodes. None of the large immature cells are capable of recirculation. At both locations, the large cells undergo a series of changes resulting in the formation of the small mature memory cell population which can be either CR+ or CR-. Each of the subpopulations formed in this maturation process can be shown to have characteristic functional attributes. The most immature large memory cell population is capable of clonal expansion when transferred to adoptive recipients in the absence of antigen challenge. In the presence of antigen, clonal expansion is suppressed. This suggests that early after triggering, the memory cell precursor is stimulated to proliferate and this proliferation continues along a predetermined path unless further antigen stimulation occurs at which point the cells are converted to antibody synthesis in a form of terminal differentiation. The next step in this maturation process results in the appearance of large and medium memory cells displaying both antigen dependent and antigen independent clonal expansions. Whether a single cell population is capable of both functions is not known but it seems likely that one is dealing with a mixture of cells in which the more mature display antigen dependent expansion. The small memory cell population formed late after immunization shows only antigen dependent expansion. Concomittant with the change in cell size one also sees a change in complement receptors. Early after immunization the large cells are predominantly CR- (95 percent) while later the small cells may be either CR+ or CR-. This suggests that the memory cell precursors, which are presumably CR+, lose complement receptors following antigen triggering and that some, but not all, re-express this surface marker. Using in vitro culture, the change in surface CR reflects a change in the capacity of the memory cells to be triggered by TI and TD antigens. The immature CR- memory cells respond to both TI and TD forms of the antigen while the mature CR+ cells respond only to the TD form and the mature CR- cells respond to both. This implies that as some of the memory cells mature and re-express the CR they lose their ability to respond to the TI antigen. This is especially interesting in light of Lite and Mullen's observation that TI antigens do not induce the formation of TI reactive memory cells. One would predict that all the memory cells generated after TI antigen priming are CR+.

The nature of the antigen may effect the kinetics of memory cell maturation. Brooks reported that immunization with DNP-BGG led to the appearance of mature memory cells by 4 to 8 weeks while chicken gamma globulin did not stimulate the production of this cell population until at least 22 weeks after priming. Whether this difference in kinetics is due to epitope density or a form of environmental priming is not yet known but leaves open the question that the nature of the antigen priming may greatly influence the repertoire of memory cell subsets which may exist at any point in time after immunization.

One potential problem with all of the above studies is the use of the adoptive transfer system. In this system, lymphocyte populations are pre-selected for cell markers such as sIg, CR, FcR and size and then transferred to irradiated host. Thus it is quite possible for the cells to modulate their expression of these markers before responding to antigen. One way to avoid these problems would be the use of an in vitro model in which both the generation of antibody forming cells and clonal expansion could be determined. Brooks reported preliminary results on this type of system. Memory cells were challenged with antigen on Day 0 of culture and then re-challenged 10 to 12 later. PFC assays were employed as a measure of cell responsiveness. The major observations made in this system were: 1) Memory cell pool expansion of 4-80 fold was detected following second challenge; 2) The expansion was antigen dependent and did not appear to involve non-immune cell recruitment into the memory population; 3) Regulatory mechanisms were also stimulated as evidenced by the appearance of free hapten augmentable plaques (a putative indication of anti-idiotypic regulation) and a non-antigen specific suppressor cell population, which, when added to fresh memory cells in ratios as low as 1:10, lead to complete inhibition; 4) The Day 0 antigenic stimulation could be accomplished using either a 24 hour pulse of antigen or continuous antigen in the culture. Continuous antigen resulted in a more consistent expansion with less regulation (hapten augmentable PFC) while the antigen pulse seemed to stimulate higher levels of regulation. This technique should prove to be very useful in evaluating the functional properties of memory cell subpopulations as well as the regulatory controls which are active on these cells.

A final point considered in the workshop concerned the effect of C3 on immune responses. For years immunologists have proposed that C3 could act either directly or indirectly to modulate antibody formation and the discovery of complement receptors on many B cell subpopulations has only served to heighten this speculation. Unfortunately, the precise role of C3 in the development of both primary and secondary responses has not yet been determined. Lucaites has approached this problem in a different way, using purified C3 and secondary in vitro PFC responses. Preliminary evidence was presented showing that when C3 was added on Day 0 of culture a profound inhibition of PFC development was observed. As little as 12.5 ug C3/ml culture was adequate to effect a 30 to 50 percent inhibition. When added to the cultures at a later time (Day 5) significant suppression was still produced suggesting that the C3 is not effecting the early stages of antigen triggering or cell proliferation but instead may be interfering with the latter stages of differentiation to plasma cells. The C3 was not toxic for the cells nor did it interfere with the secretion of antibody from preformed plasma cells. The C3 did not alter the early kinetics of PFC development but did seem to delay the decline in PFC appearance on later days (Days 8-12) suggesting that the C3 may also interfere with the development of regulatory control mechanisms. Taken together, these results can be interpreted to show that serum C3 is a major homeostatic control mechanism regulating the intensity and duration of secondary humoral responses. The important questions which must be asked are: 1) Is the effect seen with all antigens; 2) Is it apparent in primary responses; 3) Is it active on all major B cell subpopulations, and 4) If the effect is upon terminal differentiation, how is this very unique effect accomplished.

ACKNOWLEDGMENTS

Upon reflection, this was a very productive workshop and raised some very important questions. I wish to take this opportunity to thank all of the participants and especially those who presented their work. The names and addresses of the participants in the workshop are as follows: (1) S. C. Riley and W. M. Kuehl, Department of Microbiology, University of Virginia School of Medicine, Charlottsville, VA 22908, (2) H. Miller, Department of Microbiology, Michigan State University, East Lansing, MI 48824, (3) H. Lite and H. Mullen, Department of Medicine, University of Missouri School of Medicine, Columbia, MO 65201, (4) M. Hobbs, Department of Microbiology, University of Iowa College of Medicine, Iowa City, IA 52242, (5) K. Brooks, Department of Microbiology, University of Iowa College of Medicine, Iowa City, IA 52242, and (6) V. Lucaites, Department of Microbiology, University of Iowa College of Medicine, Iowa City, IA 52242.

MECHANISMS OF REGULATION OF VIRUS EXPRESSION

B.C. DEL VILLANO and G.H. BUTLER
Department of Immunology, Cleveland Clinic Foundation, Cleveland, OH 44106,
USA.

INTRODUCTION

A workshop in virology was included in the program of the Midwest Autumn
Immunology Conference for the first time in 1979. The close relationship between viruses and the immune system has been recognized since the time of
Jenner. However, it has recently been shown that viruses can be much more
than just pathogens. For example, virus related cell surface antigens may
associate with normal histocompatibility antigens, and this association may be
important in immune recognition.[1] Further, endogenous leukemia viruses are
selectively expressed in a variety of differentiated cells, and may lead to
changes in normal physiological functions of these cells.[2,3]

Three major areas involving the relationship between viruses and their hosts
were discussed in this workshop. First, the expression of murine leukemia virus
(MuLV) structural molecules in normal cells was considered. Second, a new
assay system for interferon based upon MuLV plaque reduction was described.
Third, effects of antiviral drugs on influenza virus replication were presented.

Virus Host Relationships

Oncornavirus proteins are ubiquitously distributed in the tissues of inbred
and feral mice, suggesting that the distinction between oncornaviruses and
their hosts may be arbitrary. There are several lines of evidence to support
this concept. First, in AKR mice, MuLV genomes have been shown to be physically
integrated into that of the host, and inherited as mendellian traits.[4] Second,
in strain 129 mice, the thymocyte differentiation alloantigen, G_{IX}, has been
shown to be a product of the envelope (ENV) gene of an integrated but defective
MuLV. Another virus related molecule, Abelson antigen, appears to be a
differentiation antigen in several mouse strains.[6] Abelson antigen is found
on a 120 kd protein which may be a recombinant molecule between a normal
cellular protein, NCP150, and the GAG gene of MuLV-Moloney.[7] Finally, there
are many examples of virus-host relationships in which there is overlap between
viral and cellular molecules. For example, polyoma virus particles contain
cellular histones[8] and influenza virus molecules are found on the surface of
infected cells before virus particles are present.[9]

By making a conceptual distinction between viral and virus associated cellular molecules, new approaches to understanding the relationship between oncornaviruses and their hosts have become possible. In the murine system, many copies of MuLV genomes are present, but only a few infectious viruses have been isolated. The remaining viral genetic information could code for defective viruses, or could be fragments of complete genomes. The expression of these genetic elements could be under control of cellular regulatory systems, and analysis of these systems could give new insights into viral oncogenesis. The study of these phenomena requires that the biochemical relationships among MuLV related molecules present in normal tissues, tumors, or virus particles be established.[10,11]

In the MuLV system, three classes of cell surface gp70s have been described.[10,12,13] The first is G_{IX} gp70 which was identified on the surface of thymocytes of G_{IX}^+ mice such as strain 129. Second is 0 gp70 which was found on thymocytes of C57BL/6 mice. Since these mice do not express the G_{IX} antigen, this 0 gp70 is serologically distinct from the G_{IX} gp70. Further, congenic mice were preduced which expressed both classes of gp70, suggesting that they are products of unlinked genes. The third class, X gp70, was found on certain tumor cells, and was detected using mouse antisera against these tumors.

In addition to serological differences among gp70s, structural differences have been found.[10,11,12,14] The molecular size of gp70, isolated by immune precipitation from various tissues, varies such that bone marrow cell gp70 > thymocyte gp70 > sperm cell gp70. The relationship between this structural pleomorphism and the antigenic differences among gp70s remains to be determined.

Several virus related antigens are found on leukemia cells, but not on the viruses which caused the leukemia, or on leukemia cells induced by other viruses.[15] The FMR antigen found on leukemia cells induced by MuLV-Friend, MuLV-Moloney and MuLV-Rauscher is not found on leukemia cells induced by MuLV-Gross. Conversely, Gross cell surface antigen (GCSA), which is produced by cells endogenously expressing MuLV-Gross, is not associated with cells infected by the FMR group of MuLVs. Furthermore, since the typing sera for GCSA contains antibodies to gp70, p15E and p30, GCSA may represent a non-processed polyprotein procursor unable to be incorporated into the virus, but found on the cell surface. Another possible example of this is p(75), a glycosylated polyprotein precursor of the MuLV core proteins. P(75) crossreacts with MuLV, p30, p15 and p10 and is found on spontaneous AKR thymic leukemia cells.[16] In addition, the feline oncornavirus-associated membrane antigen, FOCMA, is present on lymphocytes transformed by FeLV, but is not a virus component.[7]

Bolognesi, et al.[18] proposed an oncornavirus assembly scheme which would account for the above observations. Briefly, virion envelope components migrate to the cell membrane followed by internal virion component precursors which associate with them. As precursor processing occurs, viruses mature as buds on the cell membrane. Failures in precursor processing could account for FMR and GCSA on some cells, and the absence of such defectively processed molecules on infectious viruses. In addition, recombinational events of cellular and viral genes may produce altered virus proteins either on the surface of the cell or on the virus as with the G_{IX} and Abelson antigens.

Aside from long-term genetic events, other regulatory mechanisms may be involved in the expression of virus and virus related cell surface molecules. The interferon system represents such a mechanism. Interferons are inducible cellular molecules which prevent virus replication in otherwise susceptible cells.[19] Interferons are produced in response to virus infection or other stimuli such as double stranded RNA, BCG and some antibiotics. Several types of immune responses also result in interferon production. In addition to their anti-viral activity, interferons may inhibit cell growth and may have a regulatory role in the immune response. Once interferon is produced by the cell, it leaves the cell and subsequently binds to receptors on the plasma membranes of other cells. This receptor binding stimulates the production of at least three cellular proteins which inhibit viral replication. These proteins are not active, however, until the cell is challenged with virus. Inhibition of virus replication may be the result of altering an initiation factor or breaking down viral RNA. In addition, interferon may alter cell membranes and thereby prevent the budding of enveloped viruses from the cell.

New drugs, as well as interferon, are currently being evaluated for their clinical applicability in the treatment of virus related diseases. Antiviral agents have been of great value in defining the mechanisms of virus replication, but have been of only limited use in the prevention or cure of viral diseases. Since viruses must utilize cellular biosynthetic pathways, drugs which inhibit viral replication are frequently highly toxic to uninfected cells. Thus, to develop effective antiviral drugs, careful study of the mechanisms of virus replication and the pharmacology of new drugs is required.

Workshop Presentations

In the first presentation of this workshop the primary structure of gp70s isolated from the tissues of NZB mice was compared.[20] Molecules were radio-iodinated and isolated by immune precipitation followed by SDS polyacrylamide gel electrophoresis. Next, the labeled proteins were digested with trypsin and analyzed by two dimensional fingerprinting. The data showed major structural differences between the gp70s expressed on lymphoid tissues and that expressed in the epididymal secretions. Only slight differences in peptide fingerprints were noted among gp70s of spleen cells, thymocytes and lymph node cells. In addition, structural analysis showed that a 45 kilodalton (kd) molecule in the epididymal secretions was related to gp70-epi.

In the second workshop presentation, the immunological relatedness of gp70 derived from epididymal tissue and epididymal secretions was compared with that derived from MuLV-Scripps.[21] In homologous competition radioimmune assays, [i.e., ^{125}I gp70 (MuLV-Scripps) vs. goat anti-gp70 (MuLV-Scripps)], unlabeled gp70 from epididymal secretions competed more fully than did gp70 extracted from epididymal cells using 3 M KCl. These data indicate that at least two classes of gp70s are present in the epididymis and that the secreted molecule is more closely related to viral gp70 than the cell associated molecule.

In the third workshop presentation, evidence was presented that the source of gp70-epi may be epithelial cells within the vas deferens and epididymis.[22] Epithelial cells were cultured from tryptic or collagenase tissue digests. Gp70 was demonstrated on the surface of these cells by a staphylococcal binding assay. Cells grown on coverslips were reacted with antisera to viral gp70s, normal sera or buffer, then subsequently treated with formaldehyde fixed S. aureus (Cowan I). The binding of S. aureus to antisera treated cells demonstreated the presence of cell surface gp70.

A new method for quantitation of interferons was presented by Hsu, Finke and Proffitt.[23] This technique was based on the ability of interferon to inhibit release of MuLVs from chronically infected mouse fibroblasts. Infected cells in microculture were treated with interferon for 4 to 16 hours, and free virus released into the culture fluid was subsequently measured using the XC plaque assay. Both type I and type II interferons led to a reduced yield of ecotropic MuLV. This assay was shown to have several advantages over other interferon assays. These include: the ability to screen large numbers of samples for interferon, the need for only small amounts of reagents or test materials and the measurement of a direct effect of interferon. The assay appeared to be of equivalent or better sensitivity than other assays for interferon.

Finally, the mechanism of inhibition of influenza A virus replication by rifampicin and selenocystamine were studied by Hamzehei and Ledinko.[24] Both drugs caused a marked decrease in hemagglutinin production at nontoxic concentrations (0.025 ng/ml and 0.25 ng/ml, respectively).

In summary, this workshop represented a forum in which immunologists with a common interest in the mechanisms by which an infected host regulates the expression of viruses could discuss their diverse approaches to the problem. We hope that future meetings will encourage similar interdisciplinary workshops.

ABBREVIATIONS

The following is a list of abbreviations used in this paper: MuLV - murine leukemia virus; G$_{IX}$ - a thymocyte differentiation alloantigen associated with MuLV gp70; gp70 - major envelope glycoprotein of MuLV; ENV - MuLV gene that codes for viral envelope molecules such as gp70; GAG - MuLV gene that codes for viral internal molecules such as the group specific antigen p30; kd - kilodalton; FOCMA - Feline oncornavirus associated membrane antigen.

ACKNOWLEDGEMENTS

The authors thank Ms. M. Wright for her careful preparation of the manuscrips. Supported by Public Health Service Grant #CA 25789 from the National Cancer Institute.

REFERENCES

1. Doherty, P.C., Blanden, P.V., Zinkernagel, R.M. (1976) Transplant. Rev. 29, 89.

2. Wheelock, E.F. and Toz, S.T. (1973) Adv. in Immunol. 16, 123.

3. Proffitt, M.R., Hirsch, M.S. and Black, P.H. (1977) in Autoimmunity, N. Talal, ed., Academic Press, N.Y., pp. 385-401.

4. Chattopadhyay, S., et al. (1974) Proc. Nat. Acad. Sci. (USA) 72, 906.

5. Tung, J.S., et al. (1975) J. Exp. Med. 141, 198.

6. Risser, R., Stockert, E. and Old, L.J. (1978) Proc. Nat. Acad. Sci. (USA) 75, 3918.

7. Witte, O., Rosenberg, N. and Baltimore, D. (1979) Nature 281, 396.

8. Fay, G. and Hirt, B. (1974) Cold Spring Harbor Symp. Quant. Biol. 39, 235.

9. Rifkin, D.B., Compans, R.W. and Reich, E. (1972) J. Biol. Chem. 247, 6432.

10. Del Villano, B.C., et al. (1975) J. Exp. Med. 141, 172.

11. Lerner, R.A., et al. (1976) J. Exp. Med. 143, 151.

12. Tung, J-S., et al. (1975) J. Exp. Med. 142, 518.

13. Tung, J-S., et al. (1976) J. Exp. Med. 143, 969.

14. Del Villano, B.C., Kennel, S.J. and Lerner, R.A. (1977) in Contemporary Topics in Immunobiology, Hanna, M.G. and Rapp, F. ed., Plenum Press, New York, vol. 6, 195-208.

15. Old, L.J. and Stockert, E. (1980) in Annual Reviews Reprints: Immunology 1977-1979, Weissman, I. ed., Annual Reviews Inc., Palo Alto, CA.

16. Tung, J-S., Yoshiki, T. and Fleissner, E. (1976) Cell 9, 573.

17. Sliski, A., et al. (1977) Science 196, 1336.

18. Bolognesi, D., et al. (1978) Science 199, 183.

19. Marx, J.L. (1979) Science 204, 1293.

20. Butler, G. and Del Villano, B.C. (Abstract) Eighth Annual Midwest Autumn Immunology Conference, Detroit, MI, 1979.

21. Johnson, C., Butler, G. and Del Villano, B.C. (Abstract) Eighth Annual Midwest Autumn Immunology Conference, Detroit, MI, 1979.

22. Putinski, C.L., Butler, G.H. and Del Villano, B.C. (Abstract) Eighth Annual Midwest Autumn Immunology Conference, Detroit, MI, 1979.

23. Hsu, L., Finke, J.H. and Proffitt, M.R. (Abstract) Eighth Annual Midwest Autumn Immunology Conference, Detroit, MI, 1979.

24. Hamzehei, M. and Ledinko, N. (Abstract) Eighth Annual Midwest Autumn Immunology Conference, Detroit, MI, 1979.

III
B Cell Differentiation and Activation

INTRODUCTORY REMARKS ON B CELL DIFFERENTIATION AND ACTIVATION

J. R. BATTISTO
Department of Immunology, Research Division, Cleveland Clinic Foundation,
Cleveland, Oh. 44106. Supported by USPHA Grant AI-12468.

The transformations which a precursor B cell must undergo to become a fully-differentiated antibody-secreting plasma cell are several and varied. Furthermore, the transitions from one step to another are not clear-cut but clouded with controversy.[1] Nevertheless, there appears to be some agreement among investigators that at least four major phases of differentiation can be discerned (Table 1). At the outset there are stem cells that are totally undifferentiated. These possess the potential to acquire widely different characteristics. Secondly, there are pre-B cells that are committed for immunoglobulin synthesis since IgM is found within them. These cells are apparently without surface immunoglobulins although considerable difference of opinion exists concerning this point. Cells in this category are further delineated into large and small pre-B cells. In the third stage of development the B cells which are immature and still without contact with antigen, do

TABLE 1

MAJOR PHASES OF B CELL DIFFERENTIATION

Stage	
1	Stem cells (pluripotential to differentiate)
2	Pre-B cells, large and small (contain cytoplasmic IgM but perhaps are without sIg)
3	Immature, virgin B cells (have sIg but no contact with Ag)
	Ag ↓ Ag → Plasma Cells
	Ag ↗
4	Secondary, memory B cells (have sIg and prior contact with Ag)

demonstrate surface immunoglobulin. At this point the cells that experience antigen may be driven to become either fully developed plasma cells or memory B cells (stage 4). The memory B cells apparently have the same physical characteristics as immature, virgin cells (stage 3) except that they have had contact with antigen. The mechanisms that determine whether cells in stage 3 take either the path to become plasma cells or to become memory cells have yet to be elucidated. Thus, a number of changes that are not antigen driven occur in B cells and only the terminal stages appear to come under the influence of antigen.

The identity of the differentiating signals required prior to the exposure to antigen is not clear (see Table 2). Surely these inputs will be multiple and must be experienced at widely different times during the life of the B cell. Hormonal signals presumably may be received by stem cells during and immediately after gestation and long before most foreign antigens make their appearance. Still, no clear-cut instance of a specific hormone required by differentiating B cells has been established. For this reason hormonal agents have been placed temporarily in the category of "Other Inputs."

TABLE 2

EXTERNAL STIMULI THAT AFFECT B CELLS

Known Required Inputs Regulating Differentiation-Activation	Other Inputs
Antigens	Hormones
T cell helper factor(s)	Viruses
Macrophage-derived factor(s)	Interferons
Mature B cell product(s)	Complement
T cell suppressor factor(s)	Components
Polyclonal activators (Mitogens)	Immunoglobulin F_c
	Etc.

T cell products that are helper factors have been found to be most plentifully produced upon second exposure to an antigen.[2] Thus, helper factors are likely to be received by a particular set of B cells long after the latter have had initial contact with the antigen.

A number of macrophage-produced soluble factors have been shown to be of importance to the immune response. One of these, B cell activating factor, appears operative early in the differentiative stages [3].

Mature B cells are also capable of producing molecular substances which are capable of acting upon other B cells relatively late in their differentiation. One of these is immunoglobulin G recruiting component (GRC) which is derived from B cells most plentifully during a secondary immune response [4]. GRC causes an increase in the number of cells secreting IgG antibody long after the input of T cell replacing factor. Another factor produced by B cells is cell free supernatant (CFS) that causes IgM-bearing B cells to express surface IgE as well [5].

T cell suppressor factors that are antigen specific are typically produced by cells exposed to antigen for a second time [6]. These factors are generally genetically restricted, thus requiring some identity of the major histocompatibility locus between the T cells producing the factor and those cells responding to it [7]. Suppressor factors have been identified which inhibit immunoglobulin synthesis in vitro [8]. These factors appear to work in a manner similar to that of the suppressor cells themselves by inhibiting terminal differentiation of B lymphocytes into immunoglobulin secreting plasma cells [9]. A suppressor factor that is regulatory for B cells also emanates from the macrophage [3]. Whether this suppressor originates from T cells initially is unknown at this time.

Additional agents which affect B cells are shown in Table 2. Some of these, such as mitogens, have been shown to cause B cell differentiation but whether these substances are essential in vivo is a subject of considerable debate.

The various signals which regulate B cells must be received in one fashion or another and perhaps most are experienced through cell surface structures as opposed to intracellular receptors. The B cell surface is replete with a multitude of markers (Table 3) that range from major and minor histocompatibility antigens, differentiation antigens, surface immunoglobulins, to receptors which are specific for each of the substances listed in Table 2. Many of these markers, of course, are still functionally undefined.

TABLE 3

B CELL ANTIGENS AND/OR MARKERS

Major histocompatibility antigens
Minor histocompatibility locus antigens
Differentiation antigens
Surface immunoglobulins
Receptors for:
 Polyclonal activators
 Complement components
 Fc
 Etc. (see items in Table 2)

One of the more current interests, for instance, is the identity of the receptor(s) for T cell helper factor. Several groups of workers have pointed to different surface entities as being likely candidates for receiving this important signal. The Fc receptor of B cells has been reported by Schimpl et al.[10] to interact with TRF, the B cell marker designated as Lyb-3 has been described by Huber et al.[11] to be the responsible receptor, and Battisto et al.[12] have singled out murine differentiation antigen-1, which is stimulatory in isogeneic lymphocyte interactions, as the likely receptor for TRF. Additional work in this area is obviously required to clarify the conflicting suggestions. Furthermore, identifying the receptors for other factors and activators listed in Table 2 should gain increasing momentum in the near future.

An introduction to the differentiation and activation of B cells would not be complete without mention of the ultimate products of these cells. B cells are primarily thought of as producing antibodies, but they are now being implicated as the source of several other products (Table 4). Two of these,

IgG recruiting component[4] and cell free supernatant of B cells, which is involved in surface IgE expression[5], have already been mentioned. One of the earliest non-antibody products of B cells was identified as migration inhibition factor that had been thought to be the exclusive product of T cells[13]. Perhaps the fact that B cells are a source of interferon should not be surprising since most types of cells have been implicated in its production[14,15]. On the other hand, a somewhat unusual finding is that of Asherson and coworkers[16] who have pointed to a B cell suppressor factor that arises as a consequence of hapten-specific tolerance induction. Its molecular weight is sufficiently small as to remove it completely from the antibody molecule category. Identity of the subsets of B cells that are able to make the products listed in Table 4 is under scrutiny and undoubtedly will shortly receive more deserved attention.

TABLE 4

B CELL PRODUCTS

Antibodies
Migration inhibition factor (MIF)
IgG recruiting component (GRC)
CFS for surface IgE
Suppressor factor
Interferon

This overview of B cell differentiation and activation, however brief, will hopefully serve as a framework upon which this symposium will unfold.

REFERENCES

1. Cooper, M.D. (1979) in B Lymphocytes in the Immune Response, Cooper, M.D., Mosier, D.E., Scher, I., and Vitetta, E.S. eds., Elsevier/North-Holland, Inc., New York, vol. 3, 61-112.
2. Schimpl, A., and Wecker, E. (1972) Nature New Biol. 237, 115.
3. Albrecht, R.M. (1979) in Immunologic Tolerance and Macrophage Function, Baram, P., Battisto, J.R., Pierce, C.W. eds., Elsevier/North-Holland, Inc., New York, 81.
4. Hinchman, S. and Battisto, J.R. (1977) Immunology 33, 689.
5. Urban, J.F., Ishizaka, T. and Ishizaka, K. (1978) J. Immunol. 121, 192.
6. Kapp, J.A., Pierce, C.W., Schlossman, S., Benacerraf, B. (1974) J. Exp. Med. 140, 648.

7. Kontiainen, S., Feldmann, M., (1978) J. Exp. Med. 147, 110.
8. Krakauer, R.S., Strober, W., Rippeon, D.L. and Waldmann, T.A. (1977) Science 196,56.
9. Krakauer, R.S., Waldmann, T.A., Strober, W. (1976) J. Exp. Med. 144, 662.
10. Schimpl, A., Wecker, E., Hubner, L., Hunig, T.H. and Muller, G. (1977) in Progress in Immunology III, Mandel, T.E. et al. eds., Australian Academy in Sciences, 397.
11. Huber, G., Gershon, R.K. and Cantor, H. (1977) J. Exp. Med. 145, 10.
12. Battisto, J.R., Finke, J.H. and Yen, B. (1979) Immunology 37, 623.
13. Bloom, B.R., Stoner, G., Gaffney, J., Shevach, E. and Green, I. (1975) Eur. J. Immunol. 5, 218.
14. Epstein, L.B. (1975) in Effects of Interferon on Cells, Viruses and the Immune System, Geraldes, A. Ed., Academic Press, New York, 169-182.
15. Wietzerbin, J., Stephanos, S., Falcoff, R. and Falcoff, E. (1977) Texas Rep. Biol. Med. 35, 205-211.
16. Asherson, G.L., Zembala, M., Perera, M., Mayhew, B. and Thomas, W.R. (1977) Cell. Immunol. 33, 145.

PRODUCTION AND DIFFERENTIATION OF B LYMPHOCYTES IN THE BONE MARROW

D.G. OSMOND
Department of Anatomy, McGill University, 3640 University Street, Montreal,
Quebec, Canada H3A 2B2

Throughout life in mammals the production of B lymphocytes remains closely
associated with the hemopoietic tissues. During fetal development B lymphocyte
differentiation accompanies hemopoiesis in the liver, spleen and bone marrow
while after birth the process continues uninterruptedly in the marrow, amongst
the generation of erythrocytes, granulocytes, monocytes and thrombocytes. In
mature animals, the marrow provides a system to examine the continuous
production and differentiation of primary B lymphocytes under physiological
steady state conditions in vivo. The cellular mechanisms of this process are
of interest in relation to fundamental immunological problems, including the
generation of antibody diversity, expression of differentiation genes,
mechanisms of self tolerance and immunoregulation. The present report reviews
briefly some of our work on the bone marrow of mice, concerning firstly the
production and surface membrane maturation of virgin B lymphocytes, secondly,
some general features of their precursor cells and, thirdly, a consideration
of some possible regulatory mechanisms. Working at the single cell level to
resolve the complex cellular heterogeneity of the marrow we have employed a
combination of cell markers, including ^{3}H-thymidine in nuclear DNA plus
radioautography as a marker of cell age and renewal together with cell surface
and cytoplasmic markers of lymphocyte differentiation.

HETEROGENEITY OF LYMPHOCYTE POPULATIONS IN THE BONE MARROW

Figure 1 presents a scheme of the production and life histories of lymphocytes in general, as a framework for considering the nature of lymphocyte
populations in the marrow. The genesis of immunologically virgin lymphocytes
occurs in certain primary (central) lymphoid organs. Self-perpetuating stem
cells give rise to large proliferating precursor cells which, after a number
of cell divisions, in turn give rise to non-dividing, but immature, small
lymphocytes. These cells may then circulate to secondary (peripheral) lymphoid
tissues, including the spleen and lymph nodes, in order to become functionally
mature and to perform their definitive immunological roles. Here the cells may
die after a variable, often short, life span or be mitotically activated by the
binding of non-specific mitogens or specific antigens. This may occur once or

PRODUCTION AND TURNOVER OF LYMPHOCYTES

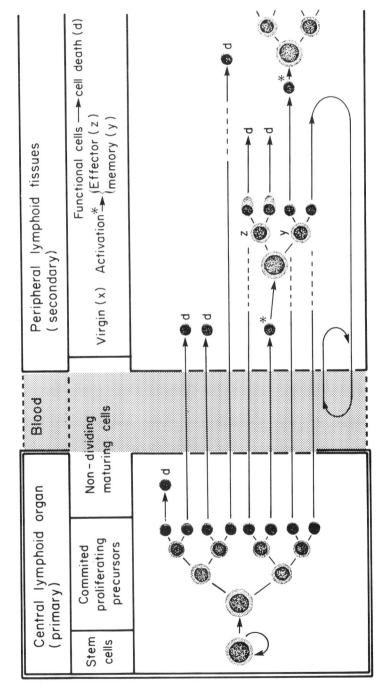

Fig. 1. Simplified general scheme of the production and the life history of lymphocytes. See text for details.

more than once in the life history of the cell and its derivatives. Primary lymphocyte production conventionally has been considered to be independent of specific antigenic stimulation. Antigen binding results in the production of short lived effector cells, together with secondary (memory) small lymphocytes, having the same antigen binding specificity as the primary cells. Secondary small lymphocytes, resulting from specific antigen binding, generally possess a long life span and have the property of recirculating continuously between the peripheral lymphoid tissues and the blood stream. This simplified outline of the lymphoid system requires modification in certain details, some of which will be considered below.

Bone marrow shows all the properties of a primary lymphoid organ, as embodied in the above scheme. There is a continuous large-scale production of small lymphocytes, an expression of B differentiation markers and function, and dissemination to the peripheral lymphoid tissues.[1] The production and turnover of marrow lymphocytes, first demonstrated more than 15 years ago,[2] have been reviewed elsewhere.[1,3-5] In general, small lymphocytes, defined by characteristic structure and cell size, comprise approximately one quarter of all the nucleated cells in the marrow of young adult mice.[6] ^{3}H-thymidine labeling and radioautography reveal that they are non-dividing cells, but most of them are rapidly renewed by the proliferation of precursor cells in the marrow.[7] After residing in the marrow parenchyma for periods ranging from a few hours to three days or so (mean marrow transit time, 14-24 hr depending upon age)[8] many of these cells can be traced by localised intramyeloid labeling to migrate to the spleen and, with a further delay, to the lymph nodes in which their accumulation and changing distribution have been studied.[4,9,10] In addition to the rapidly renewed small lymphocytes detected by ^{3}H-thymidine labeling, a substantial minority of marrow small lymphocytes remain unlabeled by ^{3}H-thymidine infusions in vivo.[8,11] This population of slowly renewing, long lived cells comprises approximately 10% of all marrow small lymphocytes in young adult mice, increasing in incidence and numbers with age.[8] Most of these cells enter the marrow from the blood stream and are part of the recirculating lymphocyte pool.[1,4,5,14] The cells include secondary B and T lymphocytes which can mediate secondary immune responses, resulting in the local development of numerous antibody producing cells in the marrow.[4,12-16] Thus, in the case of bone marrow, the distinction between primary and secondary lymphoid organs is less clear cut than depicted in Figure 1. In addition to producing and disseminating primary small lympho-cytes, the marrow receives a continuous stream of recirculating cells which after activation in the spleen can pass to the marrow in the blood stream and give rise to effector cells in the marrow parenchyma. This heterogeneity of marrow small lymphocytes has important practical consequences. In investigating marrow lymphocytes, the kinetic populations should be discriminated from one another to determine to what extent the property under investigation actually reflects that of the indigenous young primary cells rather than that of mature secondary immigrant cells.

Of the various lymphoid cell differentiation steps in the marrow, as outlined in Figure 1, the most thoroughly understood is the terminal maturation of the non-dividing small lymphocyte. Progressively earlier stages in marrow lymphocyte differentiation are progressively less clearly defined.

MATURATION OF NEWLY FORMED PRIMARY B LYMPHOCYTES IN THE BONE MARROW

In recent years, the labeling of B lymphocyte surface membrane markers has made it possible to demonstrate that much, though probably not all, of the lymphocyte production in the marrow represents the genesis of primary B lymphocytes. Surface labeling of individual small lymphocytes has been used to ask the following questions. What B cell subtypes are distinguishable in the marrow? What is the sequence of receptor expression during the critical first few days after cell production? Which developmental stages occur in the marrow, and which in the peripheral lymphoid tissues?

Marrow small lymphocytes show an unique distribution of B, T and null cells. As many as half the marrow small lymphocytes show readily detectable surface IgM molecules by radioautographic antiglobulin binding techniques, but only approximately 5% in young adult mice display high densities of Thy.1 antigen.[17] These T cells are long lived cells from the recirculating pool.[13] Virtually one half of the marrow small lymphocytes have neither readily detectable surface IgM (sIgM) molecules nor Thy.1 antigen (double negative, null cells).[17] Furthermore, the sIgM-bearing cells differ from those in the peripheral lymphoid tissues. Although in each case the IgM molecules show a relatively slow turnover in the surface membrane,[18] the density of molecules assessed by radioautographic antiglobulin binding[17] or fluorescence profiles[19,20] ranges widely from cell to cell in the marrow compared with more uniformly high densities on mature cells.

The range of density of sIgM on individual marrow small lymphocytes correlates with a comparable density range of Ia antigens expressed by the same cells.[20,21] Subgroups of these IgM and Ia antigen-bearing cells also display low to medium densities of surface IgD[22] as well as receptors for the

Fc portion of IgG and for activated complement.[23,24]   The restricted
expression of these B lymphocyte surface components, tending in each case to
be of lower density per cell than for mature B lymphocytes,[17,20,21,23]
suggest that they comprise differentiation markers in marrow lymphocytopoiesis.
H-2K antigens, in contrast, are expressed in high density on virtually all
marrow small lymphocytes, apparently regardless of their age or maturation
state.[21]

Double marker studies, combining ^{3}H-thymidine labeling in vivo with surface
rosetting and radiolabeling procedures, have established that of the marrow
small lymphocytes carrying the various B lymphocyte surface markers approximately 80% are newly formed cells, rapidly becoming labeled by
[3]H-thymidine,[21,25,26] known to be part of the indigenous, locally produced
population. These studies also verify the behaviour of B lymphocyte surface
markers as differentiation antigens on these cells, demonstrating a characteristic sequence of receptor expression during a post mitotic maturation period.
When first formed, the small lymphocytes are apparently null by surface
binding techniques. After a post mitotic lag period of approximately one day,
the cells begin to express detectable surface IgM molecules which then
increase progressively with time.[25]   Ia antigen molecules develop at the same
time as sIgM, after a comparable post mitotic lag.[21]   Only low densities of
surface IgD molecules appear on some of the rapidly renewed small lymphocytes
in the marrow (Lala, Osmond, Layton and Nossal, unpublished data) which,
together with their association with relatively high concentrations of sIgM
molecules, suggests a distinct time gap between the expression of sIgM and Ia
antigens, on the one hand, and IgD molecules on the other.

As they mature, the young marrow B small lymphocytes may be physically on
the move. The exponential renewal kinetics as well as intramyeloid labeling
studies indicate that many small lymphocytes leave the marrow while their
surface receptors are only partially developed.[1,25,26]   Receptor expression
continues after the cells enter the spleen and is probably completed as the
cells move from the red pulp into the white pulp, as indicated by homing
experiments, quantitating the development of surface markers on selectively
labeled young marrow cells recovered from the spleen after transfusion into
syngeneic recipients.[26,27,28]   Complement receptors, like IgD, are relatively
late in appearance, and develop on only approximately one half of the young B
lymphocytes.[26]   Thus, during the first 3-4 days of post mitotic maturation
the cells undergo a characteristic sequence of surface receptor expression,
being at first null, then progressively developing surface IgM, Ia antigens

and FcR followed after a further gap by CR and IgD. Although this normally occurs while the cells move from marrow to spleen, the terminal maturation of primary B lymphocytes is not dependent upon the peripheral lymphoid tissue environment but seems to be predetermined, occurring even in vitro.[25]

The sequence of receptor expression by maturing marrow small lymphocytes generally resembles the ontogenic development shown by small lymphocytes at successive stages of fetal and neonatal life.[29-32] However, there are some notable exceptions. While Ia antigens appear simultaneously with surface IgM within the adult marrow, they appear on small lymphocytes relatively late in ontogeny, considerably after the first expression of sIgM when the cells have reached the spleen.[30] The IgM-bearing small lymphocytes in neonatal spleen which have a low density of surface H-2 antigens and no Ia antigens, as detected by immunofluorescence,[30] thus appear to have no counterpart in the maturation of individual cells in post natal marrow. Whereas the ontogenic findings reflect the properties of successive generations of cells over many days or weeks, those determined by double cell labeling techniques in the adult marrow define the progressive surface membrane differentiation of individual cells over the course of the first few days after their production in the marrow.

The surface membrane development of maturing marrow lymphocytes may be examined in other ways. Small lymphocytes in the marrow have fewer surface microvilli than those in the spleen.[33] Their lectin binding pattern is also distinctive, the predominant populations in the marrow binding more wheat germ agglutinin but less concanavalin A, phytohemagglutinin-P, Soybean agglutinin, Helix pomatia lectin and Lens culinaris lectin per cell than small lymphocytes in the spleen[34] (N. Saveriano, M. Drinnan, V. Santer and D.G. Osmond, unpublished data). These findings suggest that marrow small lymphocytes show an increased number of microvilli and changes in lectin-binding surface carbohydrates, due at least in part to the development of B lymphocyte marker glycoproteins, as they circulate to the spleen.

The properties of primary B lymphocytes predicted from the kinetic and surface marker studies of marrow lymphocyte production accord well with functional assays of primary B cells. Cytotoxic doses of ^{3}H-thymidine in vivo rapidly reduce responsiveness to primary but not secondary antigenic exposure.[35] Similarly, the drug hydroxyurea, which kills cells synthesising DNA, has no immediate effect on the numbers of LPS-reactive cells in the marrow and spleen, but the numbers fall progressively during the next six to seven days.[36] These findings indicate that the responsive cells, though not themselves dividing, are newly formed from proliferating precursors. Sedimentation velocity fractionation shows that the newly formed LPS-reactive cells in marrow[37] and the hapten-specific cells responding to primary antigens in adoptive transfer assays of spleen cells of young germfree mice[38-40] are all slowly sedimenting small cells with readily detectable sIgM molecules sensitive to cytolysis by anti-IgM antibody and complement. Similarly, the period of surface membrane maturation of newly formed marrow B lymphocytes correlates with rapid functional maturation. Bone marrow itself shows relatively few antibody forming cell precursors when challenged with a primary antigen, but if held for short periods either in an antigen-free environment in vitro or in adoptive hosts such antigen responsive cells soon appear and increase in numbers.[41-44] The massive continuous production of small lymphocytes in the marrow resembles closely that of cortical thymus small lymphocytes. Like the other rapidly renewed marrow leucocytes, granulocytes and monocytes, the large scale daily production of cells must be balanced by an equal rate of cell loss under normal steady state conditions in mature animals. Kinetic evidence suggests that not only do the newly formed small lymphocytes have a short transit time in the marrow but that, in the absence of specific or non-specific activating signals, their total body life span is relatively short. Similarly, the primary responsiveness of lymph node cells identifiable by an allotypic marker show a mean half life of only a few days in host mice, suggesting that primary B lymphocytes are normally short lived cells.[45] As in the case of granulocytes and monocytes the site and nature of their death or loss are not as yet clear.

In summary, kinetic, surface marker and functional studies indicate that a large scale production of small lymphocytes in the marrow represents the formation of immature B lymphocytes that are within a few days of becoming immunologically responsive primary B cells. Their continuous emigration maintains a population of short lived primary B lymphocytes in the peripheral lymphoid tissues.

PRECURSORS OF B LYMPHOCYTES IN THE BONE MARROW

The term, pre-B cells, widely used to refer to the precursors of primary B lymphocytes, may be variously defined according to the particular experimental systems employed to characterize them. In the present discussion the term is used to include all dividing cells committed to undergo B lymphocyte different-iation that intervene between undifferentiated multipotential stem cells,[46] on the one hand, and non-dividing primary B small lymphocytes on the other

(Figs. 1, 2). In contrast, some functional assays of precursor B cell activity probably characterize only restricted parts of this cellular production pathway, while in other cases the pre-B category also includes immature non-dividing primary B small lymphocytes before the development of surface IgM and functional responsiveness, as noted below.

The rapidly renewed non-dividing small lymphocyte populations in the marrow are derived from proliferating precursors which are themselves located within the marrow. This fact, first demonstrated radioautographically in guinea pigs,[2] has been confirmed formally in mice.[7] Selective administration of ^{3}H-thymidine in vivo, designed to label DNA-synthesising cells either within or outside marrow sites respectively, excludes the possibility that the rapidly renewed small lymphocytes in the marrow originate from precursor cells located in extramyeloid tissues.[2,7,9,10] Marrow fractionation and culture techniques have identified the immediate precursors of marrow small lymphocytes as having the morphology of large lymphoid cells or transitional lymphocytes.[47] These cells differ from small lymphocytes in their larger size and leptochromatic nuclear structure but otherwise share common lymphoid features, including a minimal cytoplasmic volume and absence of cytoplasmic differentiation.[5,25,48,49] A precursor-product relationship between the marrow large lymphoid cells and small lymphocytes has been established by quantitative radioautographic studies of a wave of proliferation of partially synchronised isolated large lymphoid cells giving rise to labeled small lymphocytes in culture[47] and by visualization by time lapse photography of large lymphoid cell mitoses and their progeny.[5,50] These studies establish that at least the final generation of proliferating precursor cells which gives rise directly to non-dividing marrow small lymphocytes has the structure of large lymphoid cells.

Large lymphoid cells form a prominent cell population in mouse bone marrow, equivalent in numbers to approximately one fifth of the total marrow lymphocyte population at all ages studied.[8] They are actively proliferative. In 4 week old mice, 60% of these cells are in DNA synthesis at any one time, the proportion of DNA-synthesising cells, like their total numbers, declining progressively with age.[7] At each age the incidence of DNA-synthesising cells increases with increasing cell size, being virtually 100% in the case of the largest forms.[7] In the guinea pig, the marrow large lymphoid cells are even more numerous and more heterogeneous in degree of cytoplasmic basophilia than in the mouse.[5,50,51] Double isotope DNA-labeling studies have established that the largest cells proliferate most rapidly, with a short S phase and total cell cycle time of 3.5 hr and 6.4 hr, respectively. DNA synthesis in smaller proliferating lymphoid cells is prolonged progressively to 10.9 hr.[51]   Thus, the proliferative activity of large lymphoid cells is most active in young animals and varies according to cell size.   While it is clear that the precursors of rapidly renewed marrow small lymphocytes, and thus of the immature virgin B cells, are morphologically identifiable as large lymphoid cells, not all cells having this morphology are B cell precursors.   Other undifferentiated progenitor cells, including those which can give rise to colonies of granulocytes and monocytes in culture, also show the morphology of large lymphoid cells.[52]   Additional markers are thus necessary to identify cells in this group as being B cell precursors.

The presence and nature of surface markers on pre-B cells remain contentious. Proliferating large lymphoid cells show no sIgM molecules detectable by either radioautographic antiglobulin binding.[25]   However, approximately 60% of the large lymphoid cells do form rosettes as a result of multi point binding with anti-μ coated erythrocytes, while 20-50% can be labeled for FcR, CR and Ia antigens, including cells in DNA synthesis and some of the largest cells in the series.[26]   Thus, some B lymphocyte surface markers may be present on certain pre-B cells.   In the case of IgM it is possible that a rapid transit through the cell membrane[18] renders the molecules undetectable by surface antiglobulin binding techniques and the cells insensitive to anti-IgM antibodies in vivo, as discussed below.   On the other hand, the presence of other glycoproteins at the cell surface, as detected by lectins, may serve to characterize pre-B cells.   Current radioautographic studies of the binding of various ^{125}I-labeled lectins reveal that whereas many lectins bind to virtually all marrow small lymphocytes when used in adequate concentrations, peanut agglutinin (PNA) binds to only a small subset of these cells.   In recent double immunofluorescence labeling experiments, PNA has been found to bind to a majority of pre-B cells, defined by the presence of intracellular μ chains, while labeling only one tenth of sIgM-bearing small lymphocytes or non-lymphoid cells (D.G. Osmond and J.J.T. Owen).   Although the marker is not specific for the B cell lineage it may nevertheless prove to be useful in isolating pre-B cells.

The best documented putative marker of pre-B cells is the presence of μ-chains within the cytoplasm, in the absence of readily detectable IgM molecules at the cell surface[53,57] (M.D. Cooper, this volume).   These cells are usually detected by double immunofluorescent labeling; anti-μ-FITC binding by viable cells in suspension to reveal sIgM being followed by exposure to anti-μ-rhodamine in fixed cytocentrifuge preparations to label cytoplasmic μ chains (± sIgM). Recently, we have also labeled such cytoplasmic μ-bearing (cμ⁺) cells radioautographically after exposing fixed cells to ^{125}I-anti-μ (D.M. Rahal and D.G. Osmond, unpublished data).

The incidence of cμ⁺ cells in mouse marrow ranges in various reports from approximately 3% to 10% of all marrow nucleated cells.[53-57] In a recent study we find a mean incidence of 4-6% in CBA mice aged 7-9 weeks, equivalent to approximately half to three quarters the incidence of sIgM-bearing cells in the same experiments (D.G. Osmond and J.J.T. Owen). The variation in reported incidences may reflect differences in experimental animals, reagents and optical conditions. However, all studies agree that the cμ⁺ cells form a substantial population of cells in adult mouse marrow, exceeding the known incidence of large lymphoid cells.

We have examined the size distribution profile, morphology and kinetics of cμ⁺ cells in 8 week old CBA mice (D.G. Osmond and J.J.T. Owen). In cyto-centrifuge preparations, the cells bearing surface IgM are all small cells, measuring 5-10 μm (peak, 7.5 μm) and corresponding with previous definitions of non-dividing small lymphocytes in similar preparations based on radio-autography of DNA-synthesising cells. In contrast, cμ⁺ cells show a wide range of size, 7-15 μm in diameter. Morphologically, many of these cells resemble large lymphoid cells. In addition, the largest cells in the series exceed the size of large lymphoid cells, appearing as undifferentiated blast-like cells with plentiful cytoplasm. Such cells would not be distinguishable from blast cells of non-lymphoid cell lines without the μ chain marker. However, as many as one half of the cμ⁺ cells are small cells, measuring less than 10 μm diameter, overlapping the size distribution profile of sIgM-bearing cells and showing the characteristic morphology of small lymphocytes. By surface labeling techniques, such cells would be classified as null small lymphocytes. The present results indicate that approximately one half of the null small lymphocytes in the marrow actually have cytoplasmic μ chains, and are thus assumed to be of B lineage. The remaining null small lymphocytes have no demonstrable immunoglobulin components either within the cytoplasm or on the cell surface. They are, nevertheless, rapidly generated cells and current studies aim to determine the nature of this putative "non-B" lineage or group of cells.

Many cμ⁺ cells in mouse marrow are actively cycling, as shown by ^{3}H-thymidine labeling and radioautography of DNA-synthesising cells.[54] We have examined the cell size distribution of proliferating cμ⁺ cells by inducing metaphase arrest with vincristine sulfate for 2-4 hours in vivo (D.G. Osmond and J.J.T. Owen). The cμ⁺ cells in metaphase are all large cells, exceeding 10 μm in diameter in cytocentrifuge preparations. This finding indicates that many, if not most, of the cμ⁺ cells smaller than 10 μm diameter may be non-cycling post mitotic cells. Together with their high incidence and morphology this suggests that these are late cells in the B lymphocyte differentiation sequence, i.e., newly formed post mitotic null small lymphocytes just prior to expression of sIgM. Figure 2 illustrates this interpretation, consistent with all current data. In a continuous time sequence, cycling cμ⁺ precursors give rise to cμ⁺ small lymphocytes which subsequently express surface IgM without further cell division. The small cμ⁺ cells may tentatively best be regarded as immature primary B lymphocytes rather than pre-B cells, as defined above.

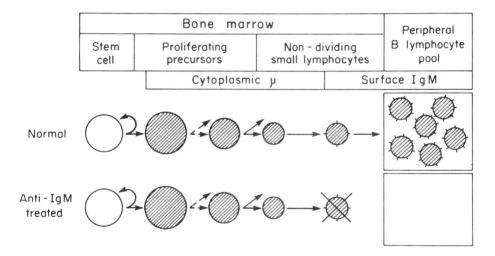

Fig. 2. Scheme of the production of B lymphocytes in the bone marrow of normal mice and the deletion of immature B lymphocytes in mice treated with anti-IgM antibodies, as detailed in the text.

The accumulation of cμ⁺ metaphase figures during vincristine sulfate treatment indicates that the proliferating forms are cycling rapidly. Preliminary data (D.G. Osmond and J.J.T. Owen) suggest that as many as 20% of the total cμ⁺ cell population enter mitosis within a 2 hour period, thus reflecting a mean apparent cell cycle time of 10 hours. If, in fact, only half the cμ⁺ cells are cycling they would be thus dividing every 5 hours on the average. This is consistent with the rapid cycling of large lymphoid cells determined by radio-autographic analyses of the rate of efflux from DNA synthesis.[51]

The capacity of cμ⁺ cells to give rise to sIgM-bearing small lymphocytes has yet to be established formally. However, such a precursor-product relationship is strongly indicated by circumstantial evidence, including the sequence of cell development in ontogeny[53,55] and during recovery from lympho-toxic drugs and anti-IgM treatment,[57] as well as the organ distribution of cμ⁺ cells and their morphological identity with large lymphoid cells.[53,54] A large to small cell sequence of successive generations of cμ⁺ cells is also indicated by studies during recovery from sublethal irradiation in mice. We have shown by quantitative ^{125}I-antiglobulin binding and radioautography that the depletion of marrow lymphocytes produced by 150 rads whole body X-irradiation is followed by the sequential appearance of various small lympho-cyte subsets, first null, then weakly sIgM-positive and finally strongly sIgM-positive cells.[58] In double immunofluorescence studies we now find that these cells are preceded at 4 days by the appearance of cμ⁺ cells that are at first entirely large cells, exceeding 10 μm diameter in cytocentrifuge preparations. Subsequently, the cμ⁺ cells increase in numbers and show a shift in size distribution profile with time to include progressively larger proportions of small cells from 4 to 10 days (D.G. Osmond and J.J.T. Owen, unpublished data). These studies not only further support the view that large cycling cμ⁺ cells give rise to smaller cμ⁺ cells and thus to sIgM-bearing small lymphocytes, but also provide a biological enrichment of marrow cμ⁺ cells. At 6 days after 150 rads whole body X-irradiation cμ⁺ cells comprise nearly 15% of all nucleated marrow cells, together with only minimal numbers of sIgM-bearing small lymphocytes (1%), making this a promising preparatory step in isolating cμ⁺ cells from marrow cell suspensions.

In general, the current findings on cμ⁺ cells are in good agreement with those of functional pre-B cell assays characterizing the properties of cells which can give rise to responsive B lymphocytes either *in vitro* or in irradiated adoptive hosts. The size distribution profile of cμ⁺ cells in cytocentrifuge preparations correlates well in shape and in its relationship to the sIgM-bearing cell profile with the sedimentation velocity profile of cμ⁺ cells.[54] The latter, ranging from 3 to 7 mm/hour,[54] in turn corresponds closely with the sedimentation velocity profile of functional pre-B cells which give rise after a maturation period to slowly sedimenting small lymphocytes reacting to lipopolysaccharide or specific antigenic stimulation by the formation of antibody producing cells.[37,59] As well as the similarity in size distribution, the susceptibility of functional pre-B cells to the action of cell cycle specific cytotoxic drugs[36] and an apparent lag of 3-4 days before large pre-B cells mature into reactive small cells[37] would all accord with the view that functional pre-B cells and $c\mu^+$ cells are overlapping if not identical populations.

Many questions concerning $c\mu^+$ cells remain unanswered, particularly concerning the earliest stages in their development. At what stage in the B lymphocyte lineage are μ chains first synthesised? What is the relationship between marrow stem cells and $c\mu^+$ cells? While multipotential stem cells are capable of repopulating the B lymphocyte system in irradiated animals, it is not known whether primary B lymphocytopoiesis is normally maintained in the steady state by the proliferation of pre-B cells or by an influx of cells from the stem cell pool.[46] In either event, how many generations of pre-B cells precede the production of primary B lymphocytes? The lack of firm data on this point is indicated in Figure 2 by a break in the depicted flow of cells. The timing of cell development in ontogeny and regenerative marrow sequences suggests that an interval of 3-4 days lapses between the appearance of the earliest pre-B ($c\mu^+$) cells and surface IgM-bearing small lymphocytes.[53-57] If so, and assuming a mean cell cycle time in the order of 10 hours, this would represent at least 6-8 cell generations. Although tentative, these considerations stress that pre-B cell development is associated with a sizeable proliferative expansion. Further clarification of these points will be relevant to the mechanisms and timing of the generation of antibody diversity in pre-B cells and the continuing genesis of novel clonotypes throughout postnatal life[60] (N. Klinman, this volume).

In summary, the precursors of newly formed small lymphocytes in the marrow have been identified among a population of highly proliferative large lymphoid (transitional) cells. The presence of cytoplasmic μ chains in the absence of readily detectable surface IgM characterizes a sizeable population of putative pre-B cells, ranging widely in size, including small lymphocytes, large lymphoid cells and large blasts. The larger cells in the series are cycling rapidly and probably include the cells defined by functional pre-B assays. It is proposed that the smaller $c\mu^+$ cells are immature primary B small lymphocytes in the post mitotic lag period prior to expression of sIgM. A large to small sequence of $c\mu^+$ cells is suggested by their recovery from X-irradiation and other lymphotoxic agents. While at least several generations of pre-B cells probably occur in the marrow, the early stages in the lineage, the precise number of cell generations and their kinetic behaviour require further study.

CONTROL OF LYMPHOCYTE PRODUCTION IN THE BONE MARROW

The continuous proliferation of pre-B cells in the marrow maintains the numbers of rapidly renewed primary B lymphocytes in the marrow and peripheral lymphoid tissues within closely predictable limits. Other marrow cell precursors which maintain peripheral cell populations of restricted life span in a similarly dynamic steady state, e.g., erythrocytes and granulocytes, are subject to delicate regulatory mechanisms mediated by a hormone-mediated negative feedback from the peripheral cell pool. Thus, variation in levels of circulating erythropoietin, acting on early erythroid precursors in the marrow, increases the rate of production of erythrocytes if the latter are withdrawn from the circulation or hemolysed, and suppresses their production if the peripheral population size is made excessive by erythrocyte hypertransfusion. While the postulated proliferative sequence of pre-B cells would permit a fine modulation of production rate, the lymphopoietic homeostatic control mechanisms are unknown. An additional type of control in the case of marrow lymphocytes is created by the need to deal with the generation of potentially self-reactive B cell clones. To address these questions we have examined marrow lymphocyte production in mice depleted of peripheral B lymphocytes by anti-IgM treatment and in mice exposed to various levels of environmental antigens.

Deletion of B lymphocytes with anti-IgM in vivo. Mice have been given repeated injections of rabbit anti-mouse IgM antibody either from birth or at 7 weeks of age after treatment with either cyclophosphamide or X-irradiation and marrow reconstitution.[61] The results of this treatment, detailed elsewhere,[62,63] are summarised schematically in Figure 2. B small lymphocytes bearing readily detectable sIgM are eliminated completely in anti-IgM treated mice, not only from peripheral lymphoid tissues such as the spleen and lymph nodes, but also from the marrow. These mice lack circulating IgM and are incapable of mounting humoral immune responses but have normal T cell function.[61]

Despite the lack of sIgM-bearing cells, the marrow shows the normal incidence of sIgM-negative small lymphocytes. Unlike the spleen, in which the large majority of residual cells are Thy.1 antigen-bearing T cells, these sIgM-negative cells are null small lymphocytes, showing no T cell markers.[62,63] However, as in normal marrow nearly half of these null cells are $c\mu^+$ cells, revealed by radioautographic labeling (D.M. Rahal, J. Gordon and D.G. Osmond, unpublished data), while ^{3}H-thymidine labeling demonstrates that virtually all of them (97%) are newly formed cells less than 3 days post mitotic age.[62]

Moreover, the marrow of anti-IgM treated mice contains large lymphoid cells[63] and a normal number of cμ⁺ cells.[57]   The first stage at which differentiating B cells become susceptible to anti-IgM antibodies in vivo thus appears to be when non-dividing cμ⁺ small B lymphocytes begin to express readily detectable sIgM molecules, one day or so after their production from proliferating pre-B cells. An actual elimination of these cells rather than a simple modulation or masking of sIgM by the injected anti-IgM antibody is suggested by the fall in total numbers of small lymphocytes in anti-IgM suppressed mice, the absence of detectable rabbit anti-mouse IgM on the surface of the residual marrow cells and the clearly pycnotic appearance of the rare sIgM-positive cell seen in the marrow. Further evidence of the identity of the anti-IgM susceptible target cell is given by the rapidity with which sIgM-positive, LPS-responsive cells appear in cultures of marrow cells from anti-IgM treated mice.[57]

The foregoing studies confirm the lack of functional surface IgM receptor molecules on pre-B cells and also suggest that the anti-IgM treated mouse may provide a model for the deletion of differentiating primary B lymphocytes by antigens, including self antigens. sIgM-negative small lymphocytes separated from mouse marrow by the fluorescence activated cell sorter (FACS) contain very few cells capable of reacting to haptens. Such reactive cells rapidly develop in tissue culture but this maturation is abrogated almost completely if cultured in the presence of haptenated human gamma globulin, acting as a B cell tolerogen.[42,43,44]   It may be concluded that newly formed primary B lymphocytes pass through a phase in which confrontation with specific antigen capable of binding to surface IgM induces tolerance. This can occur as the cells first express low densities of sIgM and Ia antigens in the marrow, before they emigrate to complete their surface receptor development and become responsive to antigenic activation in the peripheral lymphoid tissues. As in the case of anti-IgM treatment, which eliminates every immature B lymphocyte upon reaching the susceptible stage of development, exposure to a specific antigen, including self antigen, in the marrow may result in a selective elimination of the antigen-binding cells, "clonal abortion".[64]

Lack of negative feedback control of marrow lymphocyte production. The anti-IgM suppressed mouse offers an appropriate experimental system to study the regulatory effect of the peripheral B lymphocyte pool or its products on marrow lymphocytopoiesis. The selective elimination of maturing B lymphocytes and consequent depletion of the peripheral pool, without at the same time eliminating marrow precursor cells, may be likened to the induction of an hemolytic anemia in the erythroid system. If the proliferating marrow lymphocyte precursors were influenced by a feedback effect from the peripheral pool, the anti-IgM treated mice would be expected to show a compensatory increase in the rate of production of immature marrow small lymphocytes, analogous to the stimulated production of immature erythrocytes, reticulocytes, which characterises peripheral erythrocyte depletion.

To test this concept in current work[65] (G. Fulop, J. Gordon and D.G. Osmond) mice have been treated with rabbit anti-mouse IgM antibodies from birth and infused continuously with ^{3}H-thymidine for periods of 1 to 4 days after reaching 8 weeks of age. Compared with controls, treated with normal rabbit serum from birth, the residual marrow small lymphocytes in the anti-IgM treated mice showed a more rapid increase in ^{3}H-thymidine labeling index, indicating a more rapid turnover of the residual marrow small lymphocyte population (1 day: α-IgM, 39.4±0.5%; control, 24±0.6%; 2 days: α-IgM, 63.0±0.5%; control, 43.3±0.4%). This is consistent with the concept that the newly formed B lymphocytes are eliminated after a short post mitotic transit time in the marrow. However, the lymphocyte population in anti-IgM treated mice is numerically smaller than the total marrow lymphocytes of control mice. When the ^{3}H-thymidine labeling index of the marrow small lymphocytes is combined with their absolute numbers, the total rate of appearance of newly formed marrow small lymphocytes (x10^5 per femur) is indistinguishable from that in control mice (1 day: α-IgM, 7.9±0.1; control, 7.2±0.3; 2 days: α-IgM, 11.6±0.2, control, 12.1±0.1) (G. Fulop and D.G. Osmond). Thus, the total number of small lymphocytes produced in the marrow is not influenced by anti-IgM treatment and a regulatory feedback from the peripheral lymphocyte pool has not been demonstrated.

Although this finding suggests a difference in type of regulatory mechanism between marrow lymphocytes and the other lineages of blood cells produced in the marrow it accords with observations on lymphocyte production in the thymus. Multiple thymus grafts implanted in a single host continue to show active cell production despite a considerable increase in total T lymphocyte numbers[66] suggesting that primary T lymphocyte genesis is also independent of negative feedback control from the peripheral lymphocyte pool.

Amplification of marrow lymphocyte production in antigen stimulated mice. Under the conventional view of primary lymphocyte genesis the production of lymphocytes in the marrow would be expected to occur independently of antigenic stimulation. It was thus a surprise to find that marrow lymphocyte production in germfree mice, though still substantial, was reduced as compared with conventionally reared mice, suggesting that environmental antigens might play a regulatory role.[4]

In current experiments[65] (G. Fulop and D.G. Osmond) the effect on marrow lymphocyte production of administering antigens to conventionally reared mice is being explored. Ten week old C3H mice have been examined at daily intervals from 1 to 7 days after the intravenous injection of 4×10^8 sheep red blood cells (SRBC) in phosphate buffered saline. In each case a single dose of ^{3}H-thymidine was administered 24 hours before killing the mice and the labeling of marrow small lymphocytes examined radioautographically. The percentage of marrow small lymphocytes labeled by ^{3}H-thymidine increased progressively from 20%, as seen in saline injected controls, to 30% in mice given SRBC, maximal at 4-5 days after antigen administration, indicating an increased rate of small lymphocyte renewal. The total number of small lymphocytes per femoral shaft increased simultaneously, and in combination with the labeling index indicated a 3 fold increase in the absolute number of ^{3}H-thymidine labeled small lymphocytes appearing within 24 hours in the marrow. The effect was maximal by 4 days and returned to normal by the end of a week. This reveals a transient but well marked augmentation in the appearance of newly formed small lymphocytes in the marrow as a result of a single systemic dose of SRBC. Similar effects have been observed after intraperitoneal administration of SRBC. In further series of experiments, ^{3}H-thymidine labeling analyses after SRBC stimulation have been performed on mice treated with anti-IgM antibodies from birth and on nu/nu mice, lacking B lymphocytes and T lymphocytes, respectively. An increase in marrow small lymphocyte production has again been observed, comparable in magnitude with that of normal rabbit serum treated and heterozygous nu/- controls (G. Fulop and D.G. Osmond, unpublished data).

The effects of SRBC, a complex T lymphocyte dependent antigen, used at doses frequently employed in immune assays are thus apparently not restricted to the peripheral lymphoid system but also include a central effect on lymphocyte production. The mediators and cellular mechanisms of this effect are not known. Because the effect occurs in either anti-IgM treated or nu/nu mice it clearly does not require the completion of a peripheral immune response. Nor is specific binding to antigen receptors on B and T lymphocytes essential, though it cannot be excluded that binding to either B lymphocytes or T lymphocytes might be required. Although incomplete, the present evidence favours a non specific event mediated by receptors other than those binding antigen, possibly secondary to macrophage processing. Within the marrow, further experiments, currently in progress, are required to determine which stage(s) of the lymphocyte precursor pathway may be modulated by environmental stimuli, as well as the nature of the effect with regard to the kinetics of cycling pre-B cells.

Regardless of the mechanism of action, the present findings raise the possibility that environmental antigens continuously exert a non-specific stimulatory effect on a basal rate of pre-B proliferation in the marrow, thus playing a regulatory role in maintaining the high rate of B lymphocyte genesis seen in conventionally reared animals. An amplifying effect has also been attributed to environmental antigens acting on young primary B lymphocytes after they reach the peripheral lymphoid tissues.[38,39,40] It has been proposed that newly formed primary B lymphocytes normally undergo a non-specific proliferation and numerical expansion in the spleen before specific antigen binding leads to a second proliferative step and the production of antibody forming cells. The primary B lymphocytes which respond to non-specific stimuli in the spleen are small, non-dividing, sIgM-positive cells.[40] Because the elimination of cells at this stage of development from the marrow by anti-IgM treatment does not abrogate the stimulant effects of antigen on marrow lymphocyte production it can be concluded that there are two separate non-specific effects of antigens: B lymphocytes may undergo a proliferative amplification by non-specific environmental stimuli both at the pre-B cell stage in the marrow and as primary B lymphocytes in the peripheral lymphoid tissues. These processes may result simply in an expansion of the size of B lymphocyte clones of predetermined specificity or may also play a role in the generation of B lymphocyte diversity.[67] (N. Klinman, this volume)

In summary, the production of lymphocytes in the marrow appears to be remarkably independent of feedback control from the peripheral B lymphocyte pool, but may be amplified by exposure to environmental antigens, apparently by a non specific mechanism. The role of this phenomenon in generating the normal repertoire of B lymphocytes as well as the homeostatic mechanisms controlling the basic level and age-related changes in marrow B lymphocyte production remain to be determined.

ACKNOWLEDGMENTS

Recent collaborative studies referred to in this review include work on cytoplasmic μ-bearing cells with Professor J.J.T. Owen at the Medical Research Council Immunobiology Unit, Department of Anatomy, University of Birmingham,

England; and on anti-IgM treated mice with Dr. J. Gordon, Department of Surgery, McGill University, Montreal, Canada. Ms. G. Fulop and Mr. D.M. Rahal contributed to many recent experiments, as noted in the text. The technical assistance of Ms. J. Michel is gratefully acknowledged. The work was supported by the Medical Research Council of Canada.

REFERENCES

1. Osmond, D.G. (1975) J. Reticuloendothel. Soc. 17, 97-112.
2. Osmond, D.G. and Everett, N.B. (1964) Blood, 23, 1-17.
3. Osmond, D.G. (1979) in B Lymphocytes in the Immune Response. Cooper, M.D., Mosier, D., Scher, I. and Vitetta, E.S. ed., Elsevier North Holland, Amsterdam, pp. 63-70.
4. Osmond, D.G. (1980) Monogr. Allergy, 16, 157-172.
5. Rosse, C. (1976) Int. Rev. Cytol. 45, 155-290.
6. Miller, S.C. and Osmond, D.G. (1974) Exp. Haematol. 2, 227-236.
7. Miller, S.C., Kaiserman, M. and Osmond, D.G. (1978) Experientia 34, 129-131.
8. Miller, S.C. and Osmond, D.G. (1975) Cell Tissue Kinet. 8, 97-110.
9. Brahim, F. and Osmond, D.G. (1970) Anat. Rec. 168, 139-159.
10. Brahim, F. and Osmond, D.G. (1976) Clin. exp. Immunol. 24, 515-526.
11. Röpke, C. and Everett, N.B. (1973) Cell Tissue Kinet. 6, 499-507.
12. Röpke, C., Hougen, H.P. and Everett, N.B. (1975) Cell. Immunol. 15, 82-93.
13. Press, O.W., Rosse, C. and Clagett, J. (1977) Cell Immunol. 33, 114-124.
14. Röpke, C. and Everett, N.B. (1974) Cell Tissue Kinet. 7, 137-150.
15. Benner, R., van Oudenaren, A. and de Ruiter, H. (1977) Cell. Immunol. 34, 125-137.
16. Rozing, J., Brons, N.H.C. and Benner, R. (1978) Immunology 34, 909-917.
17. Osmond, D.G. and Nossal, G.J.V. (1974) Cell. Immunol. 13, 117-131.
18. Melchers, F., von Boehmer, H. and Phillips, R.A. (1975) Transplant. Rev. 25, 26-58.
19. Scher, I., Sharrow, S.O., Wistar, R., Jr., Asofsky, R. and Paul, W.E. (1976) J. exp. Med. 144, 494-506.
20. Lala, P.K., Johnson, G.R., Battye, F.L. and Nossal, G.J.V. (1979) J. Immunol. 122, 334-341.
21. Osmond, D.G. and Rahal, M.D. (1978) Anat. Rec. 190, 497-498.
22. Lala, P.K., Layton, J.E. and Nossal, G.J.V. (1979) Eur. J. Immunol. 9, 39-44.
23. Yang, W.C. and Osmond, D.G. (1979) J. Immunol. Methods 25, 211-225.
24. Chan, F.P.H. and Osmond, D.G. (1979) Cell. Immunol. 47, 366-377.
25. Osmond, D.G. and Nossal, G.J.V. (1974) Cell. Immunol. 13, 132-145.
26. Yang, W.C., Miller, S.C. and Osmond, D.G. (1978) J. exp. Med. 148, 1251-1270.
27. Ryser, J.-E. and Vassalli, P. (1974) J. Immunol. 113, 719-728.
28. Yoshida, Y. and Osmond, D.G. (1978) Transplantation 25, 246-251.
29. Nossal, G.J.V. and Pike, B.L. (1973) Immunology 25, 33-45.
30. Kearney, J.F., Cooper, M.D., Klein, J., Abney, E.R., Parkhouse, R.M.E. and Lawton, A.R. (1977) J. exp. Med. 146, 297-301.
31. Rosenberg, Y.J. and Parish, C.R. (1977) J. Immunol. 118, 612-617.
32. Owen, J.J.T. (1979) Int. Rev. Biochem. 22, 1-28.
33. Kammerer, W.A. and Osmond, D.G. (1978) Anat. Rec. 192, 423-434.
34. Osmond, D.G., Santer, V. and Saveriano, N. (1979) Exp. Hematol. 7, Suppl. 6, 129.
35. Strober, S. (1975) Transplant. Rev. 24, 84-112.
36. Rusthoven, J.J. and Phillips, R.A. (1979) Proc. Can. Fed. Biol. Soc. 22. 67.
37. Lau, C.Y., Melchers, F., Miller, R.G. and Phillips, R.A. (1979) J. Immunol. 122, 1273-1277.

38. Nossal, G.J.V., Shortman, K., Howard, M. and Pike, B.L. (1977) Immunol. Rev. 37, 187-209.
39. Shortman, K., Howard, M.C. and Baker, J.A. (1978) J. Immunol. 121, 2060-2065.
40. Shortman, K. and Howard, M. (1979) in B Lymphocytes in the Immune Response. Cooper, M.D., Mosier, D., Scher, I. and Vitetta, E.S. ed. Elsevier North Holland, Amsterdam, pp. 97-106.
41. Stocker, J.W., Osmond, D.G. and Nossal, G.J.V. (1974) Immunology 27, 795-806.
42. Stocker, J.W. (1977) Immunology 32, 283-290.
43. Nossal, G.J.V. and Pike, B.L. (1978) J. exp. Med. 148, 1161-1170.
44. Pike, B.L. and Nossal, G.J.V. (1980) Eur. J. Immunol. (in press).
45. Elson, C.J., Jablonska, K.F. and Taylor, R.B. (1976) Eur. J. Immunol. 6, 634-638.
46. Abramson, S., Miller, R.G. and Phillips, R.A. (1977) J. exp. Med. 145, 1567-1579.
47. Yoshida, Y. and Osmond, D.G. (1971) Blood 37, 73-86.
48. Yoffey, J.M., Hudson, G. and Osmond, D.G. (1965) J. Anat. 99, 841-860.
49. Yoffey, J.M. and Courtice, F.C. (1970) Lymphatics, Lymph and the Lympho-myeloid Complex. Academic Press, London and New York.
50. Osmond, D.G., Miller, S.C. and Yoshida, Y. (1973) in Haemopoietic Stem Cells. CIBA Foundation Symposium 13, Associated Scientific Publishers, Amsterdam, pp. 131-156.
51. Miller, S.C. and Osmond, D.G. (1973) Cell Tissue Kinet. 6, 259-269.
52. Dicke, K.A., Van Noord, M.J. and van Bekkum, D.W. (1973) Exp. Hemat. 1, 36-45.
53. Raff, M.C., Megson, M., Owen, J.J.T. and Cooper, M.D. (1976) Nature 259, 224-226.
54. Owen, J.J.T., Jordan, R.K., Robinson, J.H., Singh, U. and Willcox, H.N.A. (1977) Cold Spring Harbor Symp. Quant. Biol. XLI, 129-137.
55. Owen, J.J.T., Wright, D.E., Habu, S., Raff, M.C. and Cooper, M.D. (1977) J. Immunol. 118, 2067-2072.
57. Burrows, P.D., Kearney, J.F., Lawton, A.R. and Cooper, M.D. (1978). J. Immunol. 120, 1526-1531.
58. Osmond, D.G. and Evoy, K. (1977) Anat. Rec. 187, 672.
59. Lafleur, L., Miller, R.G. and Phillips, R.A. (1972) J. exp. Med. 135, 1363-1374.
60. Metcalf, E.S., and Klinman, N.R. (1977) J. Immunol. 118, 2111-2116.
61. Gordon, J. (1979) J. Immunol. Methods 25, 227-238.
62. Osmond, D.G. and Gordon, J. (1979) Cell. Immunol. 42, 188-193.
63. Osmond, D.G. and Gordon, J. (1979) in Experimental Hematology Today. Baum, S.J. and Ledney, G.D. ed. Springer Verlag, New York, pp. 129-138.
64. Nossal, G.J.V., Pike, B.L., Teale, J.M., Layton, J.E., Kay, T.W. and Battye, F.L. (1979) Immunol. Rev. 43, 185-216.
65. Fulop, G. (1979) Proc. Can. Fed. Biol. Soc. 22, 103.
66. Metcalf, D. (1966) in The Thymus; Recent Results in Cancer Research 5, pp. 17-27 & 33.
67. Cunningham, A.J. (1979) in B Lymphocytes in the Immune Response. Cooper, M.D., Mosier, D.E., Scher, I. and Vitetta, E.S. ed. Elsevier North Holland, Amsterdam, pp. 167-171 and 187-201.

(See DISCUSSION on following page)

DISCUSSION OF DR. OSMOND'S PRESENTATION

Frost: Dr. Osmond you mentioned that a mature mouse generates approximately 10^8 lymphocytes each day. Do you have an idea where these 10^8 lymphocytes, which are replaced daily, die?

Osmond: No. Clearly, if 10^8 cells are being produced everyday and the animal is in the steady state, that number must be dying everyday. All the evidence suggests that they die after a relatively short life span unless they are triggered by the appropriate antigen or other stimuli in the periphery. The problem as to where and how they die is, I think, quite comparable to that of other cells like granulocytes and monocytes which also have limited life spans of just 2 or 3 days. It's puzzling, if one looks at the various tissues one isn't conscious of seeing large numbers of granulocytes or monocytes dying but clearly they do. I don't think it is a problem which is unique to the small B lymphocyte in the marrow. The same problem, of course, exactly relates to the fate of the newly formed thymocyte.

Battisto: Dr. Osmond you've shown there is no overlap in the appearance of the Fc receptor and the complement receptor. Would you elaborate on that, please?

Osmond: The data show firstly that there are some rapidly produced cells in the marrow which are Fc receptor positive but not complement receptor positive. Conversely, as the cells mature, either in homing experiments or in in vivo turnover experiments, we find consistently that only about half of the newly formed marrow lymphocytes develop complement receptors. About half show Ig, Fc and complement, the other half Ig and Fc receptors, only suggesting that there is some dichotomy of virgin B cells depending on complement receptors, I have no other data to interpret it functionally.

Wegmann: I was fascinated by your observation that bone marrow cells will respond with proliferation upon addition of an antigen. It is a very interesting observation and one would presume they are not plaque forming cells in the bone marrow, is that correct?

Osmond: Yes, that is correct.

Wegmann: One thing I might suggest is perhaps that cells must proliferate before they undergo clonal abortion. Have you thought of this as a possibility? I'm not sure how this would be shown. At least these are interesting questions about mechanisms of tolerance in the bone marrow.

Osmond: Well, certainly it is of interest to us that a dose of sheep red cells which is commonly used in in vivo immunological experimentation is producing a marked effect in this central organ. Still, it doesn't seem to be working as an antigen. I think that is the important thing. In anti-mu suppressed mice there are no antigen binding cells and yet the sheep red cells still produce this effect. So it is not acting as an antigen and I think in comparison with work of Ken Shortman, it may be that quite non-specific stimulants, like mineral oil and things of this sort, may prove to be equally effective. Whether proliferation is a necessary part of clonal deletion is an interesting idea. Based on our work with anti-mu, anti-IgM, we would be inclined to say, if that represents anything of a model, that it's the emerging non-dividing M cell which is susceptible to deletion.

Cooper: There are cells of the monocyte-macrophage series derived from
               cells in the bone marrow that have the morphological appearance
               of small lymphocytes. It could be that some of the null cells
               that you are seeing are of that lineage. If so, one might
               expect that they might express Ia. Have you looked for that
               marker on that population of cells, as well as for the Fc
               receptor.

Osmond: We have not looked for the coincidence of Ia and Fc receptors.
               I take the point that some of these null small lymphocytes,
               particularly those with Fc receptors, might be of the monocytic
               lineage or some other parallel lineage. We are separating
               cells with a view to answering that question, at the moment.

Stavitsky: It is my impression that bone marrow does participate in the
               secondary immune response.

Osmond: Yes, very much so.

Stavitsky: How would you account for the difference between bone marrow
               not mounting a primary response but mounting a secondary
               response?

Osmond: Well, the reason, I think, that the bone marrow doesn't mount
               a primary response is that when the cells are still within
               the marrow, they are young cells that haven't yet become fully
               responsive. They don't have a full surface membrane component
               of receptors. Their IgD is still very deficient and so on.
               It isn't until they reach the spleen that they become fully
               developed. So while in the marrow they are not capable of
               responding. But if marrow is held in an antigen free
               environment for a day or two, so the cells can't get out,
               then responsive cells do appear within a day or two. So
               it would seem that the cells in the marrow are within a day or

two of becoming primary responsive cells. But _in vivo_ they
are not in that state in the marrow. On the other hand, the
secondary response has been studied in great detail by
Robert Benner in Rotterdam and his work shows very clearly that
the marrow in a secondary immune response produces antibody
and has more plasma cells actually than any other organ in
the body. This is mediated by cells that are activated in the
spleen which then travel to the bone marrow in the recirculating
pool. He has used a combination of splenectomy and parabiosis
experiments to make this point.

ONTOGENY OF THE CELLULAR EXPRESSION OF IMMUNOGLOBULIN GENES

Max D. Cooper, William E. Gathings, Alexander R. Lawton, and John F. Kearney
Cellular Immunobiology Unit, Lurleen B. Wallace Tumor Institute, Departments of
Microbiology and Pediatrics, University of Alabama in Birmingham, Birmingham,
Alabama 35294

INTRODUCTION

The study of early events in B cell differentiation has been a primary focus
of several laboratories, including ours, over the past several years. There
are two notable reasons for this preoccupation with the ontogeny of B cells.
In the first place, most of the important decisions with regard to expression
of immunoglobulin genes must take place very early in B cell development.
Second, many of the clinically relevant abnormalities in antibody producing
cells probably reflect defects evident very early in B cell differentiation.
In this report we attempt to summarize available information on some of the
more interesting features that characterize early B cell differentiation. The
approach will be to interpolate information gained from studies of normal B
cell ontogeny in several vertebrate species. In addition, certain points will
be drawn from examples of abnormal B cell differentiation.

THE SITES OF INDUCTION OF B CELL DIFFERENTIATION DIFFER IN BIRDS, MAMMALS AND
AMPHIBIANS

One prerequisite for the study of cellular aspects of early B cell differentiation was to determine the sites in which this process begins. In studies
in the chicken, early removal of the hindgut lymphopoietic organ, the bursa of
Fabricius, was shown to inhibit development of antibody producing cells[1].
Embryonic removal of the bursa can result in complete agammaglobulinemia[2].
This finding coupled with direct observations on the ontogeny of B cells
indicates that the bursal epithelial environment provides the unique inductive
signals which trigger migratory stem cells to begin lymphoid differentiation
and immunoglobulin synthesis[3]. After expansion of these clones of immature B
cells by replication, they are seeded to tissues throughout the body[4].

In mammals, the sites of B cell generation appear to differ according to
age[5-11]. During embryonic development of mice, rabbits and humans, B

lymphocytes expressing surface immunoglobulin (IgM) are generated within the liver. The results of one study suggest that stem cells are commited to differentiation along B cell lines even before they migrate to the fetal liver[12]. Later in life B lymphocytes are generated in the bone marrow.

Studies in an amphibian, Rana pipiens, suggest that B cell generation begins in the larval pronephros and mesonephros[13].

THE PATTERNS OF IMMUNOGLOBULIN EXPRESSION BY THE FIRST RECOGNIZABLE CELLS OF B LINEAGE DIFFER IN BIRDS AND MAMMALS

The first immunoglobulin producing cells to appear in the avian bursa express IgM both within the cytoplasm and on the cell surface[4]. Because IgM molecules are expressed on the cell surface and because these early B cells are exquisitely sensitive to cross-linkage of surface IgM, embryonic treatment with anti-μ antibodies completely aborts B cell development in the chick[14-16]. A surprisingly different pattern of immunoglobulin expression occurs in the earliest cells of B lineage in mammals. The first clue for this came from evidence of IgM biosynthesis by mouse fetal liver cells several days before surface IgM positive cells are detectable by immunofluorescence[7,8]. A rational explanation for these paradoxical findings became apparent with the identification in fetal liver of cells that express μ chains in their cytoplasm but not on the surface[17]. By immunofluorescence staining with fluorochrome-labeled anti-μ, they exhibit a distinctive perinuclear and reticular staining pattern of low intensity[9,17]. Cells with this unusual pattern of immunoglobulin expression later appear in bone marrow where they can be found throughout life[18]. Pre-B cells, as these cells are called, are seldom found in other tissues of the body[9]. They lack characteristic B cell surface components, including Fc and C3 receptors and, in mice, Ia determinants[19,20]. Large rapidly-dividing pre-B cells give rise to progeny of similar phenotype which replicate much more slowly, if at all, before expressing the cell surface IgM characteristic of immature B lymphocytes (see chapter by Osmond). Evidence supporting the idea that pre-B cells are the precursors of B lymphocytes can be summarized as follows: (i) pre-B cells antecede B lymphocytes during onto-geny[9,11,17]; (ii) large, rapidly-dividing pre-B cells are found only in tissues capable of generating B lymphocytes[9,22]; (iii) pulse labeling studies with ^{3}H-thymidine show that large pre-B cells immediately incorporate this DNA label which later appears in small pre-B cells and thereafter in sIgM[+] B lymphocytes (22; see chapter by Osmond in this volume); (iv) malignant clones of B cells in

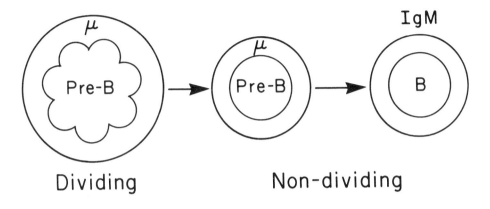

Fig. 1. Precursor-progeny relationship of pre-B and B cells.

"transitional" forms of acute lymphocytic leukemia and multiple myeloma include cells with the morphologic and immunofluorescent staining characteristics of both pre-B and B cells (23, our unpublished observations); and (v) nitrosourea- and Abelson virus-transformed cell lines with pre-B cell characteristics can be induced to become surface IgM+ B lymphocytes[24-26].

Cellular expression of immunoglobulin at the cytoplasmic level several days before surface immunoglobulin receptors can be visualized by immunofluorescence has the important implication that antigens do not play a role in the initial generation of clonal diversity. In this context it should be mentioned that some evidence has been presented for the expression of surface IgM receptors by pre-B cells[7,27]. In addition to immunofluorescence evidence, however, two sets of data appear to convincingly refute this. (i) Although treatment with cross-linking antibodies to surface IgM determinants completely aborts the development of sIgM+ B lymphocytes, this does not affect the development of pre-B cells in fetal liver or bone marrow[21]. (ii) The presence of surface IgM components on pre-B cells could not be confirmed by biochemical analysis. Specifically, the use of immunochemical methods capable of demonstrating sur- face IgM monomers on as few as 40,000 immature B lymphocytes failed to reveal surface IgM molecules on more than 300,000 pre-B cells[28].

MURINE AND HUMAN PRE-B CELLS EXPRESS μ CHAINS ONLY

It was initially deduced from immunochemical and immunofluorescence evidence that murine pre-B cells synthesized both heavy (μ) and light chains[7,17,29]. Careful re-analysis of immunofluorescence staining, however, failed to reveal light chains in cytoplasmic μ+ pre-B cells in mice and humans (28,30; our unpublished observations). Moreover, light chains could not be detected in some carcinogen- and Abelson virus-induced cell lines comprised of cells resembling pre-B cells which synthesized μ chains[25,31]. Thus the possibility was raised of an asynchronous onset of μ- and light-chain synthesis during ontogeny.

Hybridoma technology was then used to obtain large numbers of cells at a pre-B cell stage for analysis of their immunoglobulin products[30]. A non-producer variant of a HAT-sensitive myeloma cell line when fused with mouse spleen cells consistently generated hybridomas producing both heavy and light chains. Fusion of this line with mouse fetal liver cells, however, generated hybridomas which synthesized μ chains in abundance, but no light chains. The μ chains produced by the pre-B hybridomas were always slightly larger than myeloma μ chains, expressed in the cytoplasm only, and were not secreted. This method of analysis of pre-B cell products has the distinct advantage that clones of pre-B hybridomas can be raised to provide large amounts of their μ chain products for detailed immunochemical analysis.

For several reasons it was deemed necessary to re-examine the biosynthetic capabilities of normal pre-B cells. Earlier biosynthesis studies had suggested that murine fetal liver cells, obtained before the appearance of B lymphocytes, synthesize both μ heavy chains and light chains[7,32]. It could be argued that the failure to detect light chains in μ+ pre-B cells by immunofluorescence is due to a relative insensitivity of this technique. It might also be argued that pre-B cell lines and pre-B hybridomas do not accurately reflect the biosynthetic capabilities of normal pre-B cells. Therefore, the immunoglobulin biosynthetic capabilities of fetal liver cells from mouse fetuses were re-examined[28]. These studies confirmed that normal pre-B cells synthesize μ chains but not light chains. The μ chains produced by normal pre-B cells appeared to be slightly smaller than myeloma secreted μ chains. In addition, pre-B cell free μ chains were demonstrated. The apparent size differences between μ chains produced by normal pre-B cells and those made by pre-B hybridomas and the facts that the former are secreted whereas the latter are not remain unexplained.

RABBIT PRE-B CELLS EXHIBIT ALLELIC EXCLUSION

The availability of specific antibodies to allotypic determinants on V_H and C_κ regions of rabbit imunoglobulin molecules has permitted further study of the genetic events involved in generation of antibody diversity at the pre-B cell level of differentiation. As in mice and humans, ontogenetic studies of immunoglobulin expression in rabbits have shown that cytoplasmic μ^+, surface IgM⁻ pre-B cells appear several days before sIgM⁺ B lymphocytes in fetal liver. Pre-B cells are also formed in rabbit bone marrow after hemopoiesis shifts to this location[11].

Individual plasma cells produce only one of the alternative heavy and light chain alleles in heterozygous rabbits[33,34]. Although the mechanism of the allelic exclusion phenomenon remains conjectural, it was of interest to determine if selective expression of a family of immunoglobulin genes on one of the parental chromosomes occurs as early as the pre-B cell level of differentiation. This seemed especially important because of the continuing debate over whether or not B lymphocytes exhibit allelic exclusion.

In rabbits heterozygous for the V_H allotypes a2 and a3, bone marrow pre-B cells were shown by immunofluorescence to contain either a2 or a3 determinants, never both[35]. In addition to showing allelic exclusion in the pre-B cell, this finding indicates that joining of V_H and $C\mu$ genes has occurred by this point in differentiation. This conclusion is also supported by the demonstration of an N-terminal amino acid sequence characteristic of V_{HIII} for the μ chain product of a mouse pre-B hybridoma (Bennett, J. C. and Kearney, J. F., unpublished observation).

More surprising, in view of the findings discussed earlier for mouse and human pre-B cells, was the observation that some rabbit pre-B cells (∽10%) express allotypic determinants of kappa light chains. Moreover, in rabbits heterozygous for the C_κ allotypes b4 and b5, pre-B cells express only one of the alternative alleles, never both. Allotype suppression of b^4b^5 heterozygotes by alloantibodies to one kappa chain allotype selectively inhibits development of B cells expressing that allotype but has no discernible effect on pre-B cells expressing kappa chains of the suppressed allotype[36].

INITIAL SELECTION OF IMMUNOGLOBULIN GENES MAY BE RELATED TO THE SIZE OF THE V GENE FAMILIES

Virtually nothing is known about the mechanism involved in the initial selection of immunoglobulin genes to be expressed by a cell beginning differentiation along the B cell pathway. In addition to selection of a particular

V_H gene for expression with one of the J genes and the $C\mu$ gene of the heavy chain family, the cell must decide between the kappa and lambda gene families present on other chromosomes. The mouse has been shown to have many more V genes in the kappa gene family than in the lambda gene family (i.e., approximately 300 to 1) and about the same numbers of J and C genes in these families (see chapters by Hood and Seidman in the volume). In contrast with results reported earlier[37], we have observed that the frequency of immature $sIgM^+$ B cells in the mouse that express kappa chains is also much higher than the frequency of lambda positive B cells (Table 1). This suggests that the initial selection of immunoglobulin genes is in part governed by the number of V genes in the three immunoglobulin gene families.

TABLE 1

B LYMPHOCYTES FROM NEWBORN AND ADULT CBA MICE EXHIBIT SIMILAR κ/λ RATIOS

Age	Tissue*	% $s\mu^+$ Cells Expressing:**	
		kappa	lambda
Newborn	Liver	97	3
"	Spleen	97	4
Adult	Bone Marrow	98	1
"	Spleen	98	2

*Tissues were pooled from 10 newborn mice and from 3 adult mice.
**Viable cells were co-stained with fluorescein-labeled anti-μ antibodies and either anti-κ or anti-λ antibodies conjugated with rhodamine. Similar relative frequencies were observed when B cells were co-stained for surface κ and λ determinants.

If this interpretation is correct, one would predict that selection of one or the other of the parent sets of immunoglobulin genes would occur with approximately equal frequency. On first inspection, the consistent predominance of plasma cells expressing one heavy chain or light chain allotype in heterozgous rabbits would appear to be in conflict with this prediction[38]. For example, in a^2a^3 rabbits $a3^+$ plasma cells outnumber $a2^+$ plasma cells. At the pre-B cell level, however, the picture is different. In a^2a^3 heterozygous rabbits, $a2^+$ and $a3^+$ pre-B cells are present in approximately equal frequency[35]. The allelic ratios for surface immunoglobulin on immature

B lymphocytes in heterozygous rabbits are intermediate to the values for pre-B and plasma cells. Thus it appears that B cells expressing immunoglobulins of certain allotypes are preferentially triggered to proliferate and differentiate. The basis for this remains conjectural.

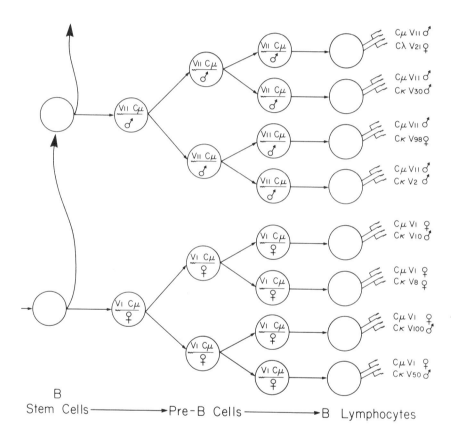

Fig. 2. Hypothetical model of the generation of B cell diversity.

GENERATION OF CLONAL DIVERSITY MAY OCCUR AT TWO LEVELS OF DIFFERENTIATON
    The model shown in Figure 2 illustrates how the generation of clonal diversity might occur. Within the proper inductive microenvironment, B stem cells

are presumed to divide and give rise to two different daughter cells, one becoming a pre-B cell on selection of a V_H gene for translocation and expression in conjunction with a J and the Cμ genes and the other retaining stem cell characteristics. This view is consistent with evidence suggesting that a single stem cell may give rise to multiple B cell clones[4]. Each daughter pre-B cell can give rise to multiple pre-B cell progeny, all of which would be expected to express the same V_H gene because the palindromal sequences thought to stablize the V-J joining appear to be excised after translocation (see chapter by Seidman). The V_H gene selected could, of course, be either a maternal or paternal gene. Also shown in the diagram is the idea that V_H genes may be randomly selected. Although it is difficult to envision a governing mechanism, it is also possible that one exists for an orderly pattern of V_H gene selection in daughter pre-B cells in view of evidence suggesting that during ontogeny B cell clones are generated in a predetermined sequence[4,39].

The model in Figure 2 also illustrates a mechanism which ensures effective clonal diversification by combined expression of the selected V_H gene with multiple V_L genes. This mechanism was proposed by Burrows, LeJeune and Kearney[30] in view of the asynchronous onset of expression of μ and light chain genes. The temporal asynchrony is clear cut in mice and humans; it also probably occurs during rabbit pre-B cell development, since light chain allotypes appear to be expressed only in the subpopulation of smaller pre-B cells (unpublished observations). If this hypothetical model is correct, the V_L gene selection process must surely be random. On the basis of evidence discussed earlier, predominant expression of $V_κ$ genes is illustrated in the model.

THE FINAL GENETIC DECISIONS IN THE ONTOGENY OF B CELL CLONES ARE CONCERNED WITH THE INTRACLONAL GENERATION OF ISOTYPE DIVERSITY

The ontogenetic order of appearance of B lymphocytes expressing different immunoglobulin isotypes is the same as the apparent phylogenetic sequence of emergence of different immunoglobulin classes. During the development of chickens, mice, rabbits and humans, sIgM[+] B cells are followed by development of sIgG[+] cells and then sIgA[+] cells[9,11,15,40]. B cells capable of synthesis of different classes of immunoglobulins are also seeded from the bursa in the same order[2,15]. Suppression of the development of sIgM[+] B cells by anti-μ treatment of chickens and mice also inhibits development of cells capable of synthesizing other immunoglobulin classes[14,15,41,42]. Consistent with this and other cellular evidence for intraclonal generation of isotype diversity is the demonstration that members of a single clone of B cells may produce immuno-globulin molecules with the same light chains and V_H region of the heavy chain but different constant regions[42,43].

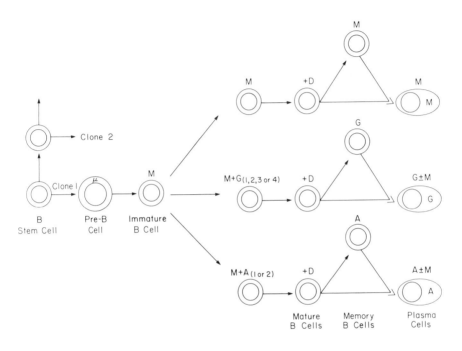

Fig. 3. Development of immunoglobulin isotype diversity within a model clone of B cells.

The model illustrating our view of the intraclonal generation of isotype diversity at a cellular level is shown in Figure 3. It incorporates some of the salient features of the ontogeny of isotype diversity. Two sets of observations seem especially noteworthy. (i) Mammalian B cells expressing one of the IgG subclasses rarely express another IgG subclass or IgA[40]. Similarly, cells expressing one IgA subclass do not express the other (Conley, M. A., unpublished observations). (ii) The first B cells that express an IgG or an IgA subclass always co-express sIgM; a few days later in development virtually all sIgG+ and sIgA+ B cells co-express both IgM and IgD[40]. These observations have important implications. The first is that the most common switch pathways are from IgM to one of the IgG subclasses and from IgM to one of the IgA sub-classes, or to IgE. The switch from IgG to IgA appears to be uncommon[44]. The second point is that B cells must be able to express simultaneously three different isotypes[9,40,45] and these are acquired in the following sequence: (i) sIgM, (ii) sIgM + sIgG or sIgA, (iii) sIgM + sIgG or sIgA + sIgD. This implies that the cells must either express multiple V_H gene copies or have an unusual mechanism for the use of long lived messengers since these cells appear destined to continue IgG or IgA synthesis for their lifetime. The ultimate solution of this and other issues raised in this discussion will require further molecular analysis of this complex cellular differentiation process. Clearly, some of the most interesting questions about the generation of B cell diversity have still not been answered.

ACKNOWLEDGMENTS

We thank Mrs. Ann Brookshire for help in preparing this manuscript, and Mrs. Susie Grey for photography and illustrations. Studies cited from our laboratory were supported in part by USPHS grants CA 16673; AI 11502; AI 14782; 5M01-RR32, and National Foundation, March of Dimes grant 1-608. Dr. Kearney is recipient of a Research Career Development Award, AI 00338.

REFERENCES

1. Glick, B., Chang, T.S. and Jaap, R.G. (1956) Poultry Sci. 35, 224.
2. Cooper, M.D., Cain, W.A., Van Alten, P.J. and Good, R.A. (1969) Int. Arch. Allergy Appl. Immunol. 35, 242.
3. Grossi, C.E., Lydyard, P.M. and Cooper, M.D. (1977) J. Immunol. 119, 749.
4. Lydyard, P.M., Grossi, C.E. and Cooper, M.D. (1976) J. Exp. Med. 144, 79.
5. Osmond, D.G. and Nossal, G.J.V. (1974) Cell. Immunol.13, 132.
6. Ryser, J-E. and Vassalli, P. (1974) J. Immunol. 113, 719.
7. Melchers, F., von Boehmer, H. and Phillips, R.A. (1975) Transplant. Rev. 25, 26.
8. Owen, J.J.T., Cooper, M.D. and Raff, M.C. (1974) Nature 249, 361.
9. Gathings, W.E., Lawton, A. R. and Cooper, M.D. (1977) Eur. J. Immunol. 7, 804.
10. Gupta, S., Pahwa, R., O'Reilly, R.O., Good, R.A. and Siegal, F.P. (1976) Proc. Natl. Acad. Sci. USA 73, 919.
11. Hayward, A.R., Simons, M.A., Lawton, A.R., Mage, R.G. and Cooper, M.D. (1978) J. Exp. Med. 148, 136.
12. Melchers, F. (1979) Nature 277, 219.
13. Zettergren, L., Kubagawa, H. and Cooper, M.D. (1980) In press.
14. Kincade, P.W., Lawton, A.R., Bockman, D.E. and Cooper, M.D. (1970) Proc. Natl. Acad. Sci. USA 67, 1918.
15. Kincade, P.W. and Cooper, M.D. (1973) Science. 179, 398.
16. Grossi, C.E., Lydyard, P.M. and Cooper, M.D. (1977) in Proceedings of the International Conference on Avian Immunology, Advances in Experimental Medicine and Biology, Benedict, A.A., ed., Plenum Press, New York, pp. 61.

17. Raff, M.C., Megson, M., Owen, J.J.T. and Cooper, M.D. (1976) Nature 259, 224.
18. Pearl, E.R., Vogler, L.B., Okos, A.J., Crist, W.M., Lawton, A.R. and Cooper, M.D. (1978) J. Immunol. 120, 1169.
19. Okos, A.J. and Gathings, W.E. (1977) Fed. Proc. 36, 1294A.
20. Kearney, J.F., Cooper, M.D., Klein, J., Abney, E.R., Parkhouse, R.M.E. (1977) J. Exp. Med. 146, 297.
21. Burrows, P., Kearney, J.F., Lawton, A.R. and Cooper, M.D. (1978) J. Immunol. 120, 1526.
22. Owen, J.J.T., Wright, D.E., Habu, S., Raff, M.C. and Cooper, M.D. (1977) J. Immunol. 118, 2067.
23. Kubagawa, H., Vogler, L.B., Capra, J.D., Conrad, M.E., Lawton, A.R. and Cooper, M.D. (1979) J. Exp. Med. 150, 792.
24. Paige, C.J., Kincade, P. W. and Ralph, P. (1978) J. Immunol. 121, 641.
25. Siden, E.J., Baltimore, D., Clark, D. and Rosenberg, N.E. (1979) Cell 16, 389.
26. Boss, M., Greaves, M. and Teich, N. (1979) Nature 278, 551.
27. Rosenberg, Y.J. and Parish, C.R. (1977) J. Immunol. 118, 612.
28. Levitt, D. and Cooper, M.D. (1979) Cell, in press.
29. Cooper, M.D., Kubagawa, H., Vogler, J.B., Kearney, J.F. and Lawton, A.R. (1978) in Advances in Experimental Medicine and Biology: Secretory Immunity, Vol. 107, McGhee, J.R., Mestecky, J. and Babb, J.L., eds., Plenum Press, New York, p. 9.
30. Burrows, P., LeJeune, M. and Kearney, J.F. (1979) Nature 280, 838.
31. Perry, R.A. and Kelley, D.E. (1979) Cell 18, 1333.
32. Melchers, F., Anderson, J. and Phillips, R.A. (1977) in Cold Spring Harbor Symposia on Quantitative Biology, Origins of Lymphocyte Diversity 41, Cold Spring Harbor Laboratory, Cold Spring Harbor, New York, p. 147.
33. Pernis, B., Chiappino, G., Kelus, A.S. and Gell, P.G.H. (1965) J. Exp. Med. 122, 853.
34. Cebra, J.J., Colberg, J.E. and Dray, S. (1966) J. Exp. Med. 123, 547.
35. Gathings, W.E., Cooper, M.D. and Mage, R.G. Eur. J. Immunol. Submitted.
36. Simmons, M.A., Hayward, A.R., Gathings, W.E., Lawton, A.R., Young-Cooper, G.O., Cooper, M.D. and Mage, R.G. (1980) Eur. J. Immunol. In press.
37. Haughton, G., Lanier, L.L. and Babcock, G.F. (1979) Nature 275, 154.
38. Mage, R.G. (1967) Cold Spring Harbor Symp. Quant. Biol. 32, 203.
39. Klinman, N.R. and Press, J.L. (1975) Fed. Proc. 34, 47.
40. Abney, E.R., Cooper, M.D., Kearney, J.F., Lawton, A.R. and Parkhouse, R.M.E. (1977) J. Immunol. 120, 2041.
41. Lawton, A.R., Asofsky, R., Hylton, M.B. and Cooper, M.D. (1972) J. Exp. Med. 135, 277.
42. Wang, A.C., Wilson, S.K., Hopper, J.E., Fudenberg, H.H. and Nisonoff, A. (1970) Proc. Natl. Acad. Sci. 66, 337.
43. Krueger, R.G., Fair, D.S. and Kyle, R.A. (1979) Eur. J. Immunol. 9, 602.
44. Rudders, R. A. Yakulis, V. and Heller, P. (1973) Am. J. Med. 55, 215.
45. Gandini, M. Kubagawa, H. and Lawton, A. R. (1980) Fourth International Congress of Immunology. Submitted.

(See DISCUSSION on following page)

DISCUSSION OF DR. COOPER'S PRESENTATION

Vitetta: Max, the free mu chain in the pre-B cells is exciting
             information that is well documented now. There are certain
             structurally unpleasing aspects of the free heavy chain in
             terms of solubility. I wonder whether you have considered
             the possibility it might be associated with beta-2-
             microglobulin? On the gels it might have run off and might
             not have been seen.

Cooper: We have considered that. It is an interesting suggestion. We
             will try it.

Battisto: Max, you have indicated that certain cells from
             agammaglobulinemics lack surface Ig but are able to show γ
             chains. This would suggest that such cells are able to
             receive T cell helper signals. Are you suggesting that T
             helper factor receptors are present before acquisiton of
             surface Ig?

Cooper: You are referring to the leukemias where we see the switch
             in heavy chain iso-types without surface receptors. We think
             that this implies that the T cell has nothing to do with
             signalling that initial genetic event. This was suggested
             some time ago by the fact that in nude mice that are lacking
             in T cells the number of cells that express other isotypes at
             a B lymphocyte level, not at the plasma cell level, is
             perfectly normal. We see the normal heterogeneity at that
             particular level. Moreover, if one grows fetal liver cells
             in culture, within a couple of days most of the parenchymal
             cells and hematopoietic cells die off so that one is left
             with a fairly pure population of immature B lymphocytes.

Now if one stimulates with LPS, as Ben Prentiss has done,
many of the cells will co-express other heavy chain isotypes,
or the number that do so will be increased dramatically. The
number of $m\mu$-γ-positive cells will be increased greatly, and
the $m\mu$-α positive cells will be increased 3 to 4-fold, as well.
If inhibitors of DNA metabolism are introduced the increased
appearance of these cells is not affected. Protein synthesis
must be blocked in order to do that. So the genetic decision
seems to be made long before T cells or an antigen comes in to
drive the system.

Rao: Yesterday, we heard a great deal about RNA additions,
deletions, splicings and so forth. How do you relate that
information to the ontogeny of B cells? Have these events
occurred at the stage of the pre-B cell or do they occur when
pre-B cells come in contact with antigen?

Cooper: Those decisions have to occur by the pre-B cell stage. What we
are detecting with our antibody are mu chain constant region
determinants. In the amino acid sequence analyses, in the
pre-B hybridoma, the N terminal amino acids indicated that these
were V_HIII. So they have the amino acids of the V region, as
well. In the rabbit one can also detect V_H allotype by the
a group. Staining pre B-cells of a heterozygous rabbit that
is a2, a3 with the appropriate reagents shows the cells will
only stain with one or the other: a2 or a3. Thus, not only
has there been a V_H region selected but it is only one of
the parent's V_H gene family. So those decisions we heard
about yesterday with regard to the heavy chain V genes and
translocation as well as joining have already been made

at this stage presumably by a cell just before it that
didn't express them. In the rabbit, as well, there are
light chain genes so that one sees V kappa allotypes and
they, too, are expressed in an allelically excluded fashion.
Whether or not they have the V gene on them we have no
direct evidence. One would assume that they, indeed, have
expressed the variable region gene, as well.

Calvanico: Why is only the heavy chain expressed in the pre-B cells?
Have you considered the possibility that it has some role
in the selection of its partners since there seems to be a
conformational preference between heavy and light chains.

Cooper: I haven't thought of it. It is an interesting idea. What is
difficult to reconcile is how the heavy chain would select not
its partner light chain but its partner light chain gene. That
is where I have trouble.

Calvanico: There are obviously problems, but I wonder whether the genes
may have some role in the selection process?

Cooper: The state that we are presently in allows us freedom to
speculate.

Battisto: Max, you touched upon a topic that you ought to have an
opportunity to expand upon. You indicated that the cells that
are secreting IgM, but which are devoid of surface IgM, stain
well. You suggested that perhaps the idiotypic portion of
the heavy chain could be expressed. Have you done any anti-
idiotypic staining of these pre-B cells to see whether that
portion alone is expressed on the surface of the cell.

Cooper: Well, we have and its diffiuclt to know how to interpret them.
In those myeloma patients we were staining with anti-idiotype

antibodies. We, indeed, saw staining that we considered
to be of pre-B cells. Most anti-idiotype antibodies, as
you know, are directed towards epitopes that are contributed
to by both light chain and heavy chain determinants. Thus,
a lot of anti-idiotypic reactivity is lost when the two
chains are separated. So that immediately presents diffi-
culties. The observations are that with these anti-idiotype
antibodies we do see cells that we think are pre-B cells
suggesting that they do bear idiotypic determinants within
the cells.

RECEPTOR-MEDIATED TRIGGERING OF MURINE B LYMPHOCYTES

ELLEN S. VITETTA, ELLEN PURÉ, LINDA B. BUCK, DOROTHY YUAN, AND JONATHAN W. UHR
Department of Microbiology and Immunology Graduate Program, University of Texas
Southwestern Medical School, Dallas, TX 75235

INTRODUCTION

Early studies of the ontogeny of B cells in the mouse indicated that the
first isotype to appear on the surface of B cells is IgM and that soon thereafter, the majority of IgM[+] cells acquire IgD.[1] This observation led to
numerous speculations concerning the roles of IgM and IgD in antigen-mediated
triggering of B cells.[2,3] In addition, it raised the issue of whether immunocompetent B cells always expressed both isotypes on their surface or whether
cells bearing only IgM could be triggered by antigen. In order to address these
questions, we have used two experimental approaches. The first is to determine
the function of different subsets of B lymphocytes that can be distinguished by
their surface Ig isotypes. For the most part, these experiments involve
deleting subpopulations of cells by treatment with C' and isotype specific antibody and investigating the immune function of the "negatively selected" cell
populations. The second approach involves removing either IgM or IgD from the
B cells by antibody-mediated capping and determining whether the "phenotypically
altered" cells can respond to antigen. The assumption of the latter type of
experiment is that treatment with intact antibody alone does not trigger the
cell.

Negative selection experiments[4,5]

In the adult mouse, approximately 90% of the B cells bear both IgM and IgD
whereas 5% express only IgM.[1] Using the protocol described in Fig. 1, we
designed experiments to evaluate the function of cells after deletion of subsets
bearing either IgM or IgD. Surviving cells were tested for their ability to
respond to the thymus-independent (TI) antigens, TNP-Brucella abortus (TNP-BA)
and TNP-lipopolysaccharide (TNP-LPS) and the thymus-dependent (TD) antigen,
TNP-sheep red blood cells (TNP-SRBC). When responses to the TD antigens were
analyzed, a source of carrier (SRBC) primed T helper cells was provided. As
seen in Table I, deletion of cells bearing either IgM or IgD reduced the in
vitro TD and TI responses significantly. Similar results were obtained when responses were evaluated by adoptively transferring cells to irradiated mice
(Table 2). Taken together, these results suggest that the majority of B cells
from adult mice which respond to TI or TD antigens bear both IgM and IgD on
their surfaces.

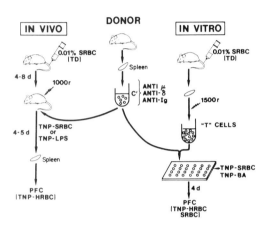

Fig. 1. Negative selection protocol for the evaluation of B cell function.

TABLE 1

PERCENT LOSS OF THE PRIMARY IN VITRO RESPONSES TO TNP-BA OR TNP-SRBC AFTER TREATMENT OF C57BL/6 SPLENOCYTES WITH C' AND RAμ AND/OR ANTI-Ig5[b].[4]

Prior treatment with C' and antiserum against	% Inhibition of response	
	TNP-BA	TNP-SRBC
δ	70	75
μ	85	90
$\mu + \delta$	93	97
Ig	90	85

Since, in adult mice the vast majority of splenic B cells bear both IgM and IgD, it was not surprising that the Ig phenotype of both the TI and TD precursors was IgM⁺IgD⁺. In contrast, in 3-6 day old neonates, >90% of the B cells bear IgM but not IgD.[6] The neonate therefore represents an attractive model for determining whether IgM⁺IgD⁻ cells can respond to TI and TD antigens. Since other studies had suggested that during ontogeny TI responses can be elicited before TD responses, it was also possible that IgM⁺IgD⁻ cells could respond to TI, but not TD antigens. Thus, the protocol described in Fig. 1

TABLE 2

PERCENT LOSS OF THE PRIMARY ADOPTIVE RESPONSES TO TNP-LPS AND TNP-SRBC AFTER
TREATMENT OF SPLENOCYTES WITH C' AND ANTI-μ OR ANTI-δ.[4]

Prior treatment with C' and antiserum against	% Inhibition of response*	
	TNP-LPS	TNP-SRBC
δ**	71	76
μ	90	85

* Adult mice; average of 8 experiments

** BDF$_1$ mice; R anti-δ

    C57BL/6 mice; hybridoma anti-δ

was repeated using spleen cells from 3-6 day old mice.[5] As seen in Table 3,
and as observed previously in experiments using cells from adult mice,[4]
treatment with C' and either anti-μ or anti-δ reduced responses to both TI and
TD antigens by greater than 70%. Thus, it was concluded that despite the fact
that less than 10% of the B cells in the neonate bear both IgM and IgD, this
subset represents the major immunocompetent cell population. Nevertheless, the
failure of anti-δ + C' to completely eliminate the TI and TD responses indicates
that IgM$^+$IgD$^-$ cells might be marginally immunocompetent. Alternatively, some
of these cells might acquire IgD during the four day culture period and then be
triggered by antigen. Regardless of the interpretation of these data, the
results support the idea that the cells which bear both IgM and IgD are most
effectively triggered by either TI or TD antigens.

Requirements for IgM and IgD for the antigen-mediated triggering of B cells[7,8]

The experiments in the preceding section defined the phenotype of the immunocompetent primary B cell but were unable to answer the question of whether one
or both receptors were required for triggering by antigen. In order to answer
this question, we performed a series of "blocking" experiments in which either
IgM or IgD was removed from cells prior to stimulation with antigen. An
important feature of the protocol used in these experiments (Fig. 2) was to
remove the Ig receptor in question both prior to and during cultivation with
antigen and T cells (in the case of a TD antigen). As shown in Fig. 3,
removal of either IgM or IgD from the cell surface completely blocked the TD

TABLE 3

EFFECT OF ANTIBODY AND C' ON THE PRIMARY IN VITRO RESPONSE OF SPLEEN CELLS FROM NEONATAL C57BL/6 MICE.[5]

AGE (days)	PRIOR TREATMENT WITH C' AND ANTISERUM AGAINST	% INHIBITION OF RESPONSE	
		TNP-BA	TNP-SRBC
3		71	88
4	IgD	77	44
6		63	78
	AVERAGE	70	70
4	IgM	72	97
6		94	98
	AVERAGE	83	97

responses to both TNP and SRBC. In contrast, the TI response to TNP-BA was blocked by treatment of cells with anti-μ but not by treatment with anti-δ. These results suggested that both IgM and IgD are required for activation by the TD antigens. In contrast, the TI antigen, TNP-BA could activate cells via IgM receptors and did not require IgD receptors. A model summarizing these results is depicted in Fig. 4. Similar results were observed in other laboratories and were extended to include a variety of other TI antigens, some of which were blocked by anti-δ and some of which were not.[9-11] Based on all these studies, it was suggested that those TI antigens which can bypass the

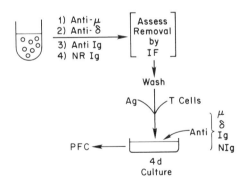

Fig. 2. Anti-Ig induced blocking of the in vitro response.

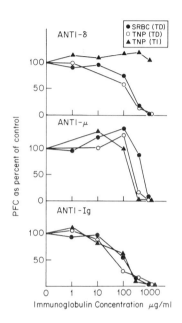

Fig. 3. Effect of antibody concentration on the inhibition of in vitro IgM responses to TD and TI antigens. Cells were treated with the indicated antiserum and GARIg under capping conditions and cultured with an Ig fraction of the same antiserum used for capping.[8]

requirement for IgD on the B cell bind to mitogen receptors[12,13] and that this interaction allows triggering of B cells in the absence of IgD. Some support for this notion was obtained by studies demonstrating that IgD and mitogen receptors were physically linked on the cell surface[14] and that some TI antigens are weak polyclonal activators.[13] Nevertheless, the possibility remains that some other feature of the antigen is involved in determining whether or not IgD receptors are required for triggering. One such possibility is the epitope density of the antigen; most TI antigens are highly multivalent while many TD antigens are paucivalent. We therefore designed experiments to determine whether there was a relationship between the epitope density of the antigen and the requirement for IgD receptors on the responding B cell.

The TNP-polyacrylamide bead system[15]

In searching for an experimental system which would allow us to investigate the role of the epitope density of an antigen in determining the requirement

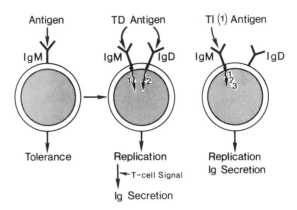

Fig. 4. Postulated role of surface immunoglobulin in delivering triggering signals to B cells.

for IgD on the B cell, two criteria were considered essential. First, the carrier portion of the antigen should be inert, i.e., neither antigenic nor mitogenic. This would facilitate the determination of the relationship between antigen structure and the requirement for IgD on the B cell in the absence of a putative mitogenic signal. Secondly, the epitope density on the carrier had to be easily manipulated. The system we chose utilized the antigen TNP-polyacrylamide (TNP-PAB), as described initially by Dintzis et al.[16] and as further evaluated by Mond et al.[17] This antigen could be rendered TI or TD by derivatizing the carrier with different numbers of TNP determinants. Moreover, it had been shown that activation of B cells is directly related to the number of epitopes on the carrier.[16,18-20] As shown in Fig. 5, polyacrylamide beads were amino ethylated and derivatized with TNP by treatment with trinitrobenzene sulfonic acid (TNBS). By varying the concentration of TNBS, antigens of different epitope densities could be synthesized. Using a radioimmunoassay to evaluate our antigen preparations, we synthesized two types of antigens, a high epitope density TNP-PAB and a low epitope density TNP-PAB, which differed in the number of epitopes present on the surface of the beads by tenfold. When spleen cells from normal mice were cultured with either one of these two antigens (Fig. 6), direct anti-TNP PFC were similar using 10^4-10^6 beads of either high or low epitope density. In contrast, underivatized beads elicited no anti-TNP response. Moreover, none of the preparations induced responses to either SRBC or horse red blood cells (HRBC). These results confirmed earlier reports that the carrier was not a polyclonal activator for murine B cells.[17]

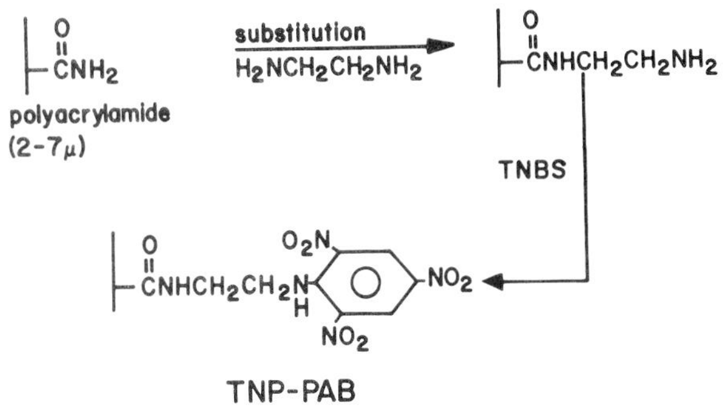

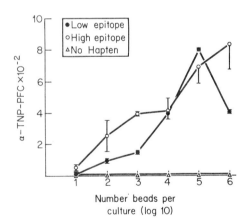

Fig. 5.   Derivatization of polyacrylamide beads with TNP.

Fig. 6.   The anti-TNP PFC response to TNP-PAB.   Spleen cells from C57BL/6 mice were cultured with varying numbers of either underivatized PAB, or low or high epitope density TNP-PAB for 4 days.   After 4 days, cells were assayed for anti-TNP PFC.   Data is presented as PFC per 10^6 viable input cells minus PFC of cultures receiving no antigen.[15]

The thymus dependency of the response to TNP-PAB. In order to determine the thymus dependency of the response to TNP-PAB, spleen cells were treated with C' and a hybridoma anti-Thy-1.2 antibody to delete T cells. This treatment effectively eliminated 30-40% of the spleen cells and, as shown in Fig. 7, abrogated the response to the low epitope density TNP-PAB. The response to the high epitope density TNP-PAB was reduced by an average of 50-60% following similar treatment. These results indicate that there is a correlation between the thymus dependency of the anti-TNP response and the epitope density of the antigen. The partial inhibition of the response to the high epitope density antigen could be due to either heterogeneity in the antigen preparation or in the responding B cells. Whatever the explanation, the results clearly indicate that at least a portion of the in vitro response to the high epitope density antigen is TI.

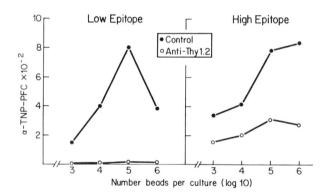

Fig. 7. Anti-TNP PFC responses to low and high epitope TNP-PAB in the absence of T cells. Spleen cells were treated with monoclonal anti-Thy-1.2 and C' or C' alone and washed extensively. Cells were then cultured with varying doses of either low epitope density (left panel) or high epitope density (right panel) TNP-PAB. After 4 days in culture cells were assayed for anti-TNP PFC. Data is presented as PFC per 10^6 viable input cells prior to T cell depletion.

IgM and IgD on B cells responding to TNP-PAB. Having established the thymus dependency of the responses to TNP-PAB, we next determined whether IgM and/or IgD were present on the responding B cells. Using the negative selection protocol described in Fig. 1, the results described in Table 4 were obtained.

TABLE 4. EFFECT OF ANTI-Ig + C' ON THE RESPONSE TO TNP-PAB

Strain		% Inhibition of the Response After Treatment with C[1] and anti-		
		δ^a	δ^b	μ
C57BL/6	Low	—	91	93
	High	—	87	88
BALB/c	Low	92	—	—
	High	98	—	—

These data confirmed the results of earlier experiments,[4,5] i.e., the B cells responding in this system also expressed IgM and IgD. In order to evaluate the requirement for IgM and IgD on B cells responding to TNP-PAB, antibody blocking experiments, analogous to those described in Fig. 2 were performed. The major modifications from the earlier system were: 1) no addition of carrier primed or "filler" T cells to the cultures and 2) the use of both a hybridoma allotype specific anti-δ and an affinity purified rabbit anti-δ. As seen in Fig. 8, the response to the low epitope density TNP-PAB was abrogated by treatment of cells with either anti-μ or anti-δ. In contrast, the response to the high epitope density TNP-PAB was completely blocked by anti-μ but only partially (30-70%) inhibited by anti-δ. These results suggest that when the epitope density of the antigen is high enough, IgD is not required on all the responding B cells. Moreover, it further confirms a correlation between the thymus dependency of the response and the requirement for IgD on the B cells. In order to explore this relationship in greater depth, we designed experiments to determine whether the TI portion of the response to the high epitope density TNP-PAB required IgD on the responding B cells. Thus, one portion of a spleen cell suspension was treated with anti-Thy-1.2 + C', a second with anti-δ, and a third with both anti-Thy-1.2 + C' and anti-δ. As shown in Fig. 9, the combined treatment reduced the response to the same extent as either deleting the T cells or blocking with anti-δ. It was therefore concluded that the TI portion of the response could be triggered in the absence of IgD receptors and, by deduction, that the TD responses to TNP-PAB requires IgD as well as IgM to activate the responding B cells.

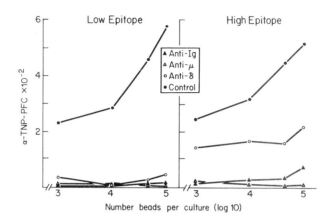

Fig. 8. Effect of anti-Ig, anti-μ, and anti-δ, on the anti-TNP-PFC response to TNP-PAB. Cells were cultured with NRIg or affinity purified anti-Ig, anti-μ, or anti-Ig5[b] antibodies for 1 hr under capping conditions. Cells were washed and 5 μg/ml of the appropriate antibody and varying doses of low (left panel) or high (right panel) epitope density TNP-PAB were added. After four days of culture cells were assayed for anti-TNP-PFC per 10^6 viable input cells minus no antigen controls.[15]

One possible explanation for these findings is that the IgM receptors on the B cells are functionally univalent whereas the IgD receptors are divalent. If we assume, as has been demonstrated by other studies, that cross-linking of receptors is a prerequisite for B cell activation,[21,22] then IgM receptors could be effectively cross-linked only by multivalent (high epitope density) antigens. IgD receptors, in contrast, would be effectively cross-linked by paucivalent (low epitope density) antigens. Thus, IgD would be required for the triggering of B cells by paucivalent antigens. The putatively different valencies of the two receptors could be related to the observation that μ chains lack a hinge region[23] while δ chains have an elongated hinge.[24] Thus, membrane bound, monomeric IgM molecules might be more "rigid", i.e., the combining sites might only bind antigen divalently if the antigenic epitopes are closely spaced on the carrier. IgD receptors, in contrast, might be more effectively cross-linked when the epitope density of the antigen is lower.

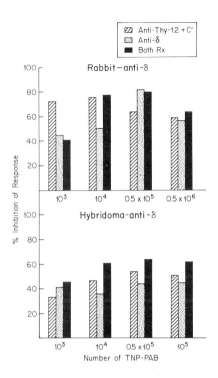

Fig. 9. Effect of anti-δ on the TI portion of the response to high epitope density TNP-PAB. Spleen cells from C57BL/6 mice were treated with either C' or anti-Thy-1.2 and C' and then washed. Cells were cultured for 1 hr in the absence or presence of rabbit anti-δ antibodies (upper panel) or hybridoma anti-δ (anti-Ig5[b]) antibodies (25 µg/ml) (lower panel) under capping conditions. After 1 hr cells were washed and resuspended in fresh medium containing antigen (high epitope density TNP-PAB) and 5 µg/ml anti-δ. Percent inhibition of the response was based on the response of cells treated with C' only. The data for hybridoma anti-δ is the average percent inhibition of four experiments.

While this explanation is consistent with the data obtained using the TNP-PAB system, it does not immediately account for two observations. First, the nature of the relationship between the thymus dependency of the response and the requirement for IgD on the B cell is not explained. It is possible that helper T cells in some way affect the ability of receptors to be cross-linked by antigen. Alternatively, IgD molecules may be involved with the reception of the T cell signal. Secondly, since all primary _in vitro_ responses thus far studied require IgM on the B cell, cross-linking _per se_ cannot adequately explain the unique requirement for interaction between IgM and antigen in triggering the B cell. It appears, therefore, that IgM receptors may deliver unique signals that IgD receptors are unable to imitate. This could be due to a specific interaction between IgM and another membrane or intracellular molecule. Another possibility is that interaction of antigen with IgM receptors alters the membrane in such a way as to effect binding of antigen or soluble factors by the IgD receptors. Clearly, the nature of the unique role of IgM remains to be explained.

Concluding Discussion

The studies described in this review offer several new insights into receptor mediated triggering of murine B cells. First, we have demonstrated that the vast majority of cells responding to both TI and TD antigens are of the IgM^+IgD^+ phenotype. Secondly, we have shown that B cells responding to TD antigens require signals mediated through both IgM and IgD receptors for subsequent differentiation into antibody secreting cells. B cell precursors responding to TI antigens can, in many instances, bypass the requirement for IgD. Finally, our studies demonstrate that the epitope density of the antigen is related to both the requirement for IgD on the B cell and the need for T cell help. The latter data has reemphasized the importance of receptor cross-linking in B cell activation and further suggests that the T cell signal is in some way related to both the epitope density of the antigen and the requirement for IgD on the B cells. With the clues provided in these studies, future experiments can be focused on dissecting the molecular associations between antigenic epitopes, B cell receptors, and T cell signals.

ACKNOWLEDGMENTS

We would like to thank Mr. Y. Chinn, Ms. M. Bagby, Ms. Y.M. Tseng, Ms. P. May, Ms. R. Summers, and Ms. J. Reed for excellent technical assistance and Ms. J. Hahn for secretarial assistance. These studies were supported by NIH Grants AI-11851, AI-12789, and CA-09082.

REFERENCES

1. Vitetta, E.S. and Uhr, J.W. (1978) Immunol. Rev. 37:50.
2. Vitetta, E.S. and Uhr, J.W. (1976) J. Immunol. 117:1579.
3. Abney, E.R., Cooper, M.D., Kearney, J.F., Lawton, A.R. and Parkhouse, R.M.E. (1978) J. Immunol. 120:2041.
4. Buck, L.B., Yuan, D. and Vitetta, E.S. (1979) J. Exp. Med. 149:987.
5. Buck, L.B. and Vitetta, E.S. (1980) Submitted for publication.
6. Goding, J.W., Scott, D.W. and Layton, J.E. (1977) Immunol. Rev. 37:152.

7. Cambier, J.C., Ligler, F.S., Uhr, J.W., Kettman, J.R. and Vitetta, E.S. (1978) Proc. Natl. Acad. Sci. USA 75:432.
8. Ligler, F.S., Cambier, J.C., Vitetta, E.S., Kettman, J.R. and Uhr, J.W. (1978) J. Immunol. 120:1139.
9. Scott, D.W., Tuttle, J. and Alexander, C. (1979) B Lymphocytes in the Immune Response. Elsevier North Holland, NY, p. 263.
10. Zitron, I.M., Mosier, D.E. and Paul, W.E. (1977) J. Exp. Med. 146:1707.
11. Mosier, D.E., Mond, J.J., Zitron, I., Scher, I. and Paul, W.E. (1977) Immune System: Genetics and Regulation. Academic Press, NY, p. 699.
12. Moreno, E. and Berman, D.T. (1979) J. Immunol. 123:2915.
13. Coutinho, A. and Moller, G. (1973) Nat. New Biol. 245:12.
14. Forni, L. and Coutinho, A. (1978) Nature (Lond.) 273:304.
15. Pure, E. and Vitetta, E.S. (1980) Submitted for publication.
16. Dintzis, H.M., Dintzis, R.Z. and Vogelstein, B. (1976) Proc. Natl. Acad. Sci. USA 73:3671.
17. Mond, J.J., Stein, K.E., Subbarao, B. and Paul, W.E. (1979) J. Immunol. 123:239.
18. Humphries, G.M.K. (1979) J. Immunol. 123:2126.
19. Feldman, M. (1972) J. Exp. Med. 135:735.
20. Desmayard, C. and Feldman, M. (1975) Eur. J. Immunol. 5:537.
21. Weiner, H.L., Moorhead, J.W. Yamaga, K. and Kubo, R.T. (1976) J. Immunol. 117:1527.
22. Sidman, C.L. and Unanue, E.R. (1979) J. Immunol. 122:406
23. Kehry, M., Sibley, C., Fuhrman, J., Schilling, J. and Hood, L.E. (1979) Proc. Natl. Acad. Sci. USA 76:2932.
24. Lin, C.L. and Putnam, F.W. (1979) Proc. Natl. Acad. Sci. USA 76:6572.

(See DISCUSSION on following page)

DISCUSSION OF DR. VITETTA'S PRESENTATION

Wegmann: Just as a speculation, I wonder if its possible that the
             T cell signal is elicited through IgD?

Vitetta: I think that this is a distinct possibility which we are
             now testing experimentally.

Klinman: There's still some confusion on the first part of your
             presentation and maybe it can most clearly be stated by
             saying - If IgM only cells are tolerized and IgM plus IgD
             cells are immunized, how then do you reduce your response by
             tolerance?

Vitetta: The cells which are tolerized may be those with a very high
             μ/δ ratio, i.e., the most immature cells. Such cells may
             fall into the μ-predominant category depending on the assay
             used to detect them.

Klinman: Is there any evidence that IgM can deliver a tolerogenic
             signal?

Vitetta: Let me just provide one answer. If you take an adult B cell
             bearing both IgM and IgD and you remove the IgD with papain or
             anti-δ antibody then those cells become more readily
             tolerizable. This may be due to the fact that the IgM
             receptors remaining on these cells don't cross-link as
             effectively with an antigen and they are thus rendered un-
             responsive by the tolerogen.

Klinman: But wouldn't the easiest explanation be that you can also
             tolerize cells bearing both IgM and IgD.

Vitetta: Yes, I think perhaps we both know that when the cell has
             very little IgD and a lot of IgM it tolerizes well.

Mullen: Could you comment on whether you think there's a distinction
             between TI-1 and TI-2 antigens with respect to the requirement
             for IgD.

Vitetta: I think the only TI-2 antigen that's been looked at adequately
             is TNP-ficoll, and the response to this antigen does in fact
             require IgD on the B cells. In most people's hands, responses
             to TI-1 antigens do not require IgD, although there are
             exceptions.

Sundick: To follow-up this interesting question of the statement that
             only the IgM bearing cell can be tolerized or easily
             tolerized - What is the evidence that the cell bearing just
             IgM without any IgD could under some circumstances produce
             an immune response, perhaps to a thymic-independent antigen?

Vitetta: I think there's good evidence that it can. As I've tried to
             say in the first slide. If you generate μ-predominant cells
             on the FACS and then adoptively transfer these cells into an
             irradiated host, they will go on and give an antibody
             response. Thus, I think that under the appropriate circum-
             stances μ-predominant cells can differentiate and respond to
             antigen. Moreover, in vivo responses to TI antigens occur
             very easily in ontogeny - before IgD develops on splenic B
             cells. All these systems have the disadvantage that the
             μ-predominant cells may differentiate and be activated at a
             later stage - such as after they acquire IgD.

Stavitsky: Could you say something briefly about the relationship of
             IgD to immunologic memory - We didn't hear this in your
             talk.

Vitetta: The only good evidence which correlates IgD with memory is evidence coming from the Herzenberg laboratory and evidence which Zan-Bar, Strober and I have which says that IgD is expressed on memory cells. I think the interesting finding is that a cell which bears only IgD (i.e., the δ-predominant cell) seems to be that cell which can propagate more memory cells. That cell will presumably go on to become an IgG secreting cell, perhaps by losing IgD from its surface and acquiring IgG. I think that's probably the extent of the evidence on that point.

Stavitsky: In the data you present you were talking exclusively about IgM in a primary response. Is there anything bearing on IgG in a primary?

Vitetta: Yes, if you adoptively transfer cells bearing IgM and IgD the cells secrete IgM followed by IgG. In the data I presented today we are looking at a four day response in vitro where we're only reading out IgM.

Battisto: I have one final question, Ellen, and that has to do with the fact that your high epitope, you say, requires T cell help or stimulates T cell help, and the question then is -

Vitetta: Half of the high epitope requires T cell help, half of it doesn't.

Battisto: OK, now the half that does - the question is since your bead, you say is inert, what's acting as the carrier in that situation?

Vitetta: We don't know the answer but the possibilities include the PAB bead (which is unlikely), the TNP itself, or the TNP-PAB bound to a self antigen on the macrophage or B cell.

Battisto: So you're saying then that the excess number of haptenic groups on the bead covers up the carrier portion.

Vitetta: I don't know what the exact explanation is. What I'm saying is that when the epitope density on the carrier is high, you don't need very much T cell help, when its low you do. The mechanism responsible for that difference is not yet clear. We've not looked at macrophages or allosteric problems with presentation of antigen - all those experiments remain to be done.

B-CELL MATURATION AND REPERTOIRE EXPRESSION

NORMAN R. KLINMAN

Department of Cellular and Developmental Immunology, Scripps Clinic & Research Foundation, 10666 North Torrey Pines Road, La Jolla, Ca. 92037

THE ORIGINS OF THE CLONOTYPE REPERTOIRE

The repertoire of unique antibody specificities (clonotypes) within individuals or inbred murine strains is exceedingly diverse.[1-5] In addition, to the extent to which it can be measured, there is considerable fidelity of genetic inheritance of this repertoire among genetically identical individuals.[5-11] The mechanism by which such a diverse repertoire can be inherited has been a major conceptual problem in the field of immunology over the past several decades. Recent findings at the level of the molecular biology of the immune system have helped enormously resolving certain elements of this issue. First, it appears that a fairly sizable repertoire is inherited.[12-14] Second, a large variety of mechanisms by which the inherited genetic information can be permuted to give rise to a substantially more diverse repertoire of messenger RNA and proteins have been defined.[12-15] Given these findings, it is no longer paradoxical that individuals may express an enormous repertoire of unique antibody specificities.

In order to account for an inherited repertoire of 10^7 to 10^8 unique specificities, several years ago we proposed a modification of germ line inheritance theories which we termed "the pre-determined permutation model" for repertoire expression.[16,17] The model was based on the assumption that a relatively limited amount of genetic information for encoding heavy and light chain variable regions could be permuted by any of a number of already proposed mechanisms such as heavy chain/light chain shuffling, as well as insertions of variable region sequences.[16-20] The recent information generated at the molecular biology level, now makes many such mechanisms, as well as other previously unforseen mechanisms highly likely. The crux of the pre-determined permutation model, however, was the proposition that there should be little or no randomness or chance in the ultimate expression of the clonotype repertoire. If, indeed, all genetically identical individuals are obligated to express during their lifetime the entire set of permutations of their inherited repertoire and, in particular, the entire set of permutations

for every inherited heavy and light chain variable region, then evolutionary
selective forces could be exerted on the inherited repertoire as a whole.
In this context random somatic events such as mutations and subsequent
environmental selective forces at the level of the maturing individual would
play little or no role in repertoire establishment. Missing from this theory
was not only a reasonable set of molecular mechanisms which could account for
such permutations, which no longer are in question, but also firm evidence
that the entire repertoire of genetically identical individuals is, in fact,
expressed identically. While mature individuals of inbred strains appear to
express identically a small set of identifiable clonotypes which predominate
responses to certain antigens,[5-11]evidence was and remains lacking for the
vast majority of the expressed repertoire.

One of the important findings which led to the predetermined permutation
postulates, was the finding that very early in neonatal life all genetically
identical individuals appeared to have a limited repertoire and to share that
repertoire to a large extent.[1,21-24] For example, during the first three days
of neonatal life all BALB/c mice share the same three predominant DNP-specific
clones as well as the same three TNP-specific clones. These findings have
been extended to show the sharing of a predominant dansyl-specific clone in
the early neonate[24] and the reproducible absence of a predominant clone
specific for fluorescence and phosphorylcholine.[23] A surprising finding was
the reproducible occurrence of PC-specific clones and the TEPC-15 clonotype
in particular towards the end of the first week of postnatal development.
Since the number of B cells is highly limited in early neonates and since
clonotypes are expressed in expanded clones representing several hundred B
cells, it is not surprising that early neonates express very few clonotypes.
Indeed, since each of the predominant early clonotypes represents approximately
one in 10^4 of neonatal B cells the early neonatal repertoire presumably
contains 10^4 clonotypes. Furthermore, since a large number of VH and VL
sequences are inherited in the fertilized gamete, it is also not surprising
that the earliest expressions of genetically identical individuals may be
highly reproducible. However, if predictions of the predetermined permutation
theory are correct, it is possible that not only will the very earliest
repertoire be inherited with great fidelity, but as new clonotypes are
expressed, these too might be expressed identically in all individuals in the
strain. If this is so, then at an intermediate period in repertoire
development, it might be expected that a diversified repertoire would be

present in each individual but that this repertoire in the strain as a whole would be limited relative to the repertoire of fully developed adults.

Experiments have been carried out using the influenza hemagglutinin specific repertoire. Adults have been shown to express an extremely diverse repertoire specific for the PR8-hemagglutinin consisting of hundreds of different clonotypes.[4,25,26] As mentioned above, the repertoire in adults specific for this antigen as well as all others tested, is too diverse to enable meaningful comparison from individual to individual. An analysis of the early neonatal repertoire specific for the PR8 hemagglutinin showed that very few B cells are responsive to this antigen. An analysis of the repertoire at two weeks of age, however, revealed quite clearly that as late as two weeks after birth the repertoire of individuals is still quite limited and that this repertoire is shared by all individuals of the strain at the same age.[27] Thus, there are only eight to ten clonotypes in two-week old BALB/c neonates which are specific for the influenza PR8-hemagglutinin and these clonotypes by and large are shared by all or most individuals of the strain at that age. Since each of these clonotypes represents one in $1-2 \times 10^6$ B cells of the two week old neonate it may be concluded that at a time when the repertoire has diversified approximately 100 fold from that seen at birth, the repertoire is reproduced with extraordinary fidelity in a genetically predetermined fashion.

It is tempting to extrapolate these findings and to suggest that since this process of diversification is so rigorous and pre-determined during its early stages, might it not continue in a similarly ordered fashion, and, thus, yield an entire specificity repertoire, expressed in a genetically predetermined fashion. Obviously, evidence for this conclusion in older animals is still lacking. Nonetheless, the findings with regard to repertoire expression during early repertoire expansion do indicate that the general principles of the predetermined permutation model appear to be correct. Given the wealth of molecular mechanisms that appear to be available for permuting the information available for variable region synthesis, these findings put enormous constraints on exactly how these mechanisms may, in fact, operate. Thus, whatever the mode for heavy and light chain combination, and DNA recombinational mechanisms, these appear to be highly ordered and entirely predictable from the inherited genetic information.

Most importantly, the finding that at least a large element of a diverse repertoire is expressed reproducibly and can be carefully defined, enables the investigator to approach perhaps the most significant questions in the area of repertoire development. Since it is apparent that all individuals inherit reproducibly the capacity to express an almost unlimited repertoire, how does the system control and select those specificities which ultimately comprise the mature functional B-cell repertoire?

MECHANISMS WHICH MAY CONTROL REPERTOIRE EXPRESSION

At the present time it may be presumed that there are at least three major influences on the expressed repertoire: A) the genetically inherited potential for variable region sequence; B) the establishment of tolerance for self-antigenic determinants, and, C) immunoregulatory phenomena, such as idiotype recognition, which may exert their influence at the level of repertoire expression. Other factors such as antigenic stimulation and suppressor T-cell function have been excluded from this list. Although antigenic stimulation or contact may affect the rate of precursor generation from the generative cell source as suggested by Dr. Dennis Osmond at this meeting, by and large, antigens appear to exert their major influence on an already mature repertoire and evidence is lacking that such contact has any influence on the repertoire other than to generate secondary B cells.[1,28] Although several investigators have implied that certain specificities appear in the repertoire only after antigenic stimulation,[29,30] careful analysis indicates that it is more likely that antigenic stimulation enables the selective expansion of certain spontaneously occurring clonotypes. Thus, after antigenic stimulation, the repertoire is skewed, in some cases, towards the disproportionate representation of rare clonotypes. A comparison of the primary and secondary B-cell repertoire in BALB/c mice specific for the influenza hemagglutinin, reveals that over 90% of secondary clonotypes have already been observed in the primary B cell pool.[4] However, certain of these clonotypes are expanded more than 100 fold in the secondary population, whereas, others are hardly expanded at all.

Similarly, suppressor T cells, as they are currently understood, appear to exert their action at the level of antigenic stimulation of the available mature B-cell repertoire. Thus, there is no current evidence that such cells affect repertoire expression per se.

A) The genetic inheritance of immunoglobulin variable regions. Using the expression of predominant clonotypes in various murine strains, a considerable amount of convincing data has been generated which establishes a linkage of the expresssion of these clonotypes to both the heavy and light chain genetic loci.[5-11,31-34] Indeed, many of the clonotypes have been mapped with respect to one another, within a locus linked to the heavy chain allotype. Overall, these data provide compelling evidence for a determinative role of genetically inherited information in the expression of, at least, the predominantly expressed clonotypic sets. However, all such linkage studies are dependent upon polymorphism in the expression of these clonotypes among strains expressing different heavy and/or light chain allotypes. Recent findings at the precursor cell level indicate that such polymorphisms may not be as absolute as they may appear at the serum level, indicating that post-maturational regulatory events may contribute to apparent polymorphisms.[35,36] Since elements of this regulation would appear to also link to the heavy chain allotype locus,[35] delineating these regulatory mechanisms from genetically inherited V regions may present an intractable problem.

In general, apparent polymorphic expressions can be tentatively grouped into three categories by examination at the precursor cell level. The first category is true polymorphisms which are best exemplified by the paucity in most murine strains of heavy chains capable of complementing the λ_1 light chain to yield anti-dextran antibody.[7,37] Certain strains such as BALB/c express these heavy chains in great abundance whereas other strains such as C57BL/6 appear to express few, if any λ bearing anti-dextran clonotypes. Category 2 might be called pseudo-polymorphism. This may include the expression of the TEPC 15 anti-PC antibody clonotype which predominates the response of BALB/c mice to phosphorylcholine.[5,8-10,35,36] This clonotype is limited, or absent in the immune serum of immunized mice of other strains. At the individual precursor cell level, however, several other strains have been found to express a clonotype which is idiotypically indistinguishable from the BALB/c clonotype.[35,36,38] While these clones are not frequent, they are expressed at a frequency equal to, or higher than the expression of most other clonotypes within these strains. The frequency of expression of this clonotype appears to be under multigenic control, one element of which is linked to the heavy chain allotype locus.

The third category may be termed partial polymorphism. Preliminary evidence may put the response to the NP-determinant in various murine strains in this category.[6,32] Preliminary work from two laboratories now indicates that the heteroclitic λ dominant response characteristic of C57BL/6 mice is also prevalant at the non-immune precursor cell level in many other murine strains in relative abundance.[39,40] The apparent polymorphism defined by heterocliticity and λ bearing antibodies is not truly observed until after extensive

immunization, which indicates that it is probably regulatory in nature.
However, although several strains appear to express λ heteroclytic clones in
abundance, these clones may not possess idiotypic identity with analogous
clones from the C57BL/6 strain.[40] Given the fact that heavy chains are the
product of at least three and possibly four separate genetic loci, such findings
may suggest that some idiotypic polymorphism may be controlled by a locus other
than that responsible for the bulk of the variable region gene. Thus, for some
clonotypic sets, polymorphism may exist in genes encoding one segment of the
variable region, but not in genes encoding others.

Future experiments at the molecular level should ultimately be able to
delineate those mechanisms which may be responsible for polymorphism at the
level of encoding immunoglobulin heavy and light chains. However, the true
contribution of inherited variable region genetic information to repertoire
expression may ultimately best be evaluated at the level of pre-B cells.
Repertoire comparisons at this level would be less likely to exhibit the
results of regulatory phenomena.

B) The role of tolerance to self-antigenic determinants. Central to
immune responsiveness is the capacity to discriminate exquisitely between self
and non-self. Indeed, the ability to retain recognition for foreign antigenic
determinants while remaining unresponsive to the myriad self determinants
may constitute the major selective pressure for diversity in the recognition
repertoire. In spite of cognizance of the central role of self-recognition in
repertoire establishment, the basic underlying mechanisms for eliminating self-
reactivity remain poorly defined. For example, it is not yet known to what
extent self determinants are not recognized by virtue of the elimination of
given reactivities at the level of genetic inheritance of variable regions
encoding self-recognizing clonotypes. Similarly, while it has been shown that
in the absence of certain self determinants, recognition of these determinants
does develop,[41] it has not yet been demonstrated that the lack of responsiveness to such determinants under normal developmental conditions is due to the
elimination of specificities at the level of repertoire expression.

Perhaps the most significant advance in this regard over the past few years
has been the clear demonstration that B cells during their development are
exquisitely susceptible to tolerance induction.[42-47] Thus, under conditions
where mature B cells are not at all affected by the presence of an antigen,
immature B cells can be totally inactivated by the mere presence of antigenic
determinants presented in a polyvalent fashion. Importantly, this mode of

tolerance induction is largely independent of the carrier used or recognition of the antigen by cells other than the B cells in question.[42] Thus, it appears that a mechanism is available whereby B cells recognizing antigenic determinants in the milieu in which they are developing, could presumably be readily eliminated from the repertoire. Whether such elimination constitutes a significant physiological mechanism for the elimination of potentially self-reactive B cells, remains a matter of speculation. Again, it would seem that a comparison of the repertoire as expressed in pre-B cells to that expressed by mature B cells will be the method of choice in establishing the extent to which tolerance plays a determinative role in repertoire expression.

   C) Immunoregulatory phenomena and repertoire establishment. Recent studies using anti-idiotype specific antibodies have demonstrated that idiotypic recognition can prohibit the stimulation of B cells bearing the iditoype in question.[48-51] In addition, it appears that idiotype specific T cells may also prohibit such stimulation.[49-51] Studies at the precursor cell level in individuals which have received anti-idiotype antibodies at an early stage in their development have demonstrated that indeed, the presence of anti-idiotypic reactivity can eliminate precusor cells bearing that idiotype from the repertoire of responsive cells.[52] Thus, it would appear that anti-idiotypic antibody like tolerogen, may be capable of influencing the repertoire, particularly with respect to the elimination of cells bearing given idiotypes during clonal maturation. While such findings give credence to the notion that so-called "network" type phenomena may impact on the B cell repertoire per se,[53,54] experiments which demonstrate this under physiological conditions, rather than artificially induced conditions, are still lacking.

   Recently, a mechanism has been described whereby immunized individuals specifically suppress the immune response of their own primary B cells to the immunizing antigen.[55] Importantly, this "antibody specific" immunoregulation, does not affect B cells allogeneic to the primed individual at the heavy chain allotype locus. Thus, in cell transfer experiments BALB/c recipients immunized to the dinitrophenyl determinant suppresses the response to dinitrophenyl of primary BALB/c B cells but will not suppress the responses of B cells of any strain differing from BALB/c in the heavy chain allotype locus. As defined, this mechanism requires antigenic stimulation to be observed and is observed at the level of stimulation of B cells already present in the mature repertoire. However, this effect is long lasting so that if during their lifetime individuals contact a multitude of antigens, one might predict that such suppression may normally accumulate for a variety of clonotypes. In addition, it is possible that if such suppression exists, then, like induced anti-idiotypic suppression, it may affect responsivness at the level of repertoire expression. Preliminary observations indicate that aged individuals exhibit a considerable amount of naturally acquired antibody specific immunoregulatory capacity, e.g: aged individuals can be shown to suppress syngeneic, but not allotype allogeneic, primary B-cell responses. Experiments are now in progress to determine whether pre-B cells or B cells in nude mice, which may lack the immunoregulatory phenomena because of the lack of T cells, in fact display clonotypes absent from the mature repertoire of conventional mice. If such differences do exist then a role for antibody specific immunoregulation would be indicated if it obtains primarily for clonotypes responsive to environmental antigens and can be displayed by transfer to allotype allogeneic recipients but not syngeneic recipients.

PROSPECTIVES IN UNDERSTANDING REPERTOIRE ESTABLISHMENT

If, as implied above, we may now assume, 1) that all individuals inherit the capacity to express in a predetermined fashion an enormous B cell repertoire, and, 2) that individuals and strains may differ mainly in repertoire expression at the level of control of which of the inherited clonotypes should and should not be expressed as mature B-cell clones, how may we perceive the underlying basis of repertoire establishment for all immunocompetent cells. The immune system is comprised of several cellular subsets, each bearing different degrees of similarity with each of the others.   Thus, among B cells, one can delineate neonatal B cells, immature adult B cells, T-independent B cells of two types, T-dependent B cells and secondary B cells; and in the T-cell population, T cells reactive in cytotoxicity assays (CTL) and those participating in proliferative or helper T-cell reactivities. It is likely that some of the B-cell subsets represent no more than progressive maturational states of the same B-cell clones, and thus, from the point of view of repertoire expression may be relatively uninteresting. On the other hand, is it possible that the neonatal B-cell repertoire is constituted of a cell line distinct from that which dominates the adult repertoire. Furthermore, since T cells may share B-cell idictypes in their surface receptors, to what extent do T-cell subpopulations represent branch points in the same cell lineage giving rise to the B-cell repertoire? Perhaps, then, information as to the mechanisms of repertoire establishment might best be acquired by comparative studies of the repertoires of the various defineable lymphoid subsets. This could enable a determination of the extent to which these repertoires may vary, by virtue of a differential genetic inheritance as opposed to differences in the selective forces imposed during their development and expression.

While defined studies of the T-cell repertoire are at a comparatively primitive stage, several phenomena are clearly displayed by the T-cell repertoire, which if present, are much more difficult to perceive within the B-cell repertoire. In particular, T cells display marked disparities in responsiveness to certain antigenic determinants which link, primarily to genes in the H-2 rather than allotype locus.[56] Although still highly controversial, recent studies of such immune response (Ir) phenomena indicate that these differences in responsiveness may be perceived as the result of the deletion of certain T-cell specificities from the repertoire by virtue of the presence of specific MHC determinants during T cell maturation or stimulation.[57,58] This is particulary clear in the case of the Fl which lack certain recogntion elements normally present in the responder parent.[58]   In addition, T-cell recognition of antigen as presented on macrophages, B cells or CTL target cells, in many instances appears to require the simultaneous recognition by the T cell of MHC determinants on the cells in question. Recently, it has been proposed (Linda Sherman and Norman Klinman, Lake Arrowhead Meeting for Immunologists, June, 1979) that, as with B cells, T cells may bear a single specific receptor and that the T-cell repertoire is also inherited as an enormous number of predetermined specificities. If this is so, then it is possible to account for skewing of the T-cell repertoire towards self-H-2 recognition, by merely imposing on the developing T-cell repertoire a severe selective pressure against clonotypes with high affinity for self MHC determinants and the preservation of the small percentage of T cells exhibiting low to intermediate affinity for self-MHC determinants. These T cells would then constitute the entire set of clonotypes reactive to environmental antigens. Such a theory places the entire burden of T-cell recognition on a single receptor much as the burden of B-cell antigen recognition resides in the single immunoglobulin receptor. This T-cell receptor would function to recognize both the antigen and the MHC of the presenting cell. Since an equivalent selective pressure is not avaialbe for B cells, the T and B-cell repertoires, though similarly derived, would rapidly diverge.

If the selective pressures on the T-cell repertoire are so obvious, does this imply that equivalent selective pressures do not obtain for the B-cell repertoire? To date, Ir-type phenomena have not been observed at the B-cell level. While certain strains may lack B-cells responsive to various antigenic determinants, it has yet to be demonstrated that F1 individuals do not display elements of the repertoire which were present in either of the two parents. It is reasonable to assume that such phenomena have not been observed simply because they have not been investigated. Thus, one may propose that the role of environmental factors, particularly those which are inherited as part of the developing milieu may best be evaluated by searching for such Ir-type phenomena and then localizing the responsible genetic locus.

Both for this type of approach, as well as an evaluation of other contributors to repertoire establishment such as immunoregulatory phenomena, it would appear that the crucial analysis is that which compares the mature, expressed repertoire to the pre-B-cell repertoire. In order to successfully carry out such analyses several criteria must be met. First, it is essential that one be able to isolate from within the bone marrow or neonatal sources, the pre-B-cell population, exclusive of any mature B cells, or B cells which have already expressed their receptors, and have thus been accessible to environmental influences. Recent experiments have demonstrated that immunoglobulin bearing cells can be eliminated from a cell suspension by either panning on anti immunoglobulin coated plates or by use of a cell sorter.[44,59-61] In addition, Dr. Harold Miller presented at this meeting, the identification of an antiserum which recognizes pre-B cells and enables them to be selectively purified by use of a cell-sorter. Thus, it has been possible to obtain reasonably pure sources of pre-B cells and begin the analysis of their specificity repertoire. It should be noted that several other approaches or analysis of the pre-B-cell repertoire are currently being pursued in other laboratories, one of which was described at this meeting by Dr. Max Cooper.

The second criterion which must be met is that the response of the pre-B cells must be carried out in a milieu which itself would not prejudice their repertoire expression. Thus, attempts to look at pre-B cells specific for murine antigens might be precluded if those murine antigens are present during the course of antigenic stimulation of these B cells. At the present time, the in vitro splenic focus technique is the method of choice for observing the response of pre-B cells.[1] In this technique, B cells are transferred intraveneously to an irradiated antigen primed recipient and, after removal of the recipient's spleen, responses are observed by stimulation of splenic fragments in culture. This technique maximizes responsiveness of all B-cell subclasses, presumably because of the excess of helper phenomena and ancilliary cells in the environment of the stimulated B cells, and apparently enables responses even of pre-B cells by allowing them to express their immunoglobulin receptor during the course of the experiment. However, such pre-B cells are developing in the milieu of a murine splenic fragment and thus may be subject to suppressive or tolerizing influences much as they would have been had they developed in their original host. Thus, judicious selection of antigens must be included if any such analysis is to provide meaningful results.

Finally, for any such analysis to be successful, the means whereby the clonotype repertoire can be delineated must be available. As described above one such system is represented by the repertoire of two week old mice specific for the influenza hemagglutinin. Because of the available panel of similar influenza hemagglutinins, it is possible to delineate the reactivity pattern of every hemagglutinin-specific clone. Thus, it has been possible to completely define the hemagglutinin-specific portion of the repertoire of fully competent and expressed B cells. If one should characterize the pre-B-cell repertoire from mice at this age or earlier, it should, theoretically be possible to determine the extent to which the expressed repertoire reflects the inherent repertoire prior to environmental contact. It is obvious that similar studies can be carried out on any number of antigen systems, many of which provide either extensive idiotypic characterization, or equally useful panels of related antigens.

Thus, it would seem that the experimental procedures are now available which could delineate many aspects of the mechanisms which may control repertoire expression. A comprehensive analysis of the pre-B-cell repertoire and a comparison of these inherited specificities among strains should soon be available. By determining the relationship of this repertoire to that expressed in mature lymphoid cells of the various T and B-cell subsets one may ultimately begin to understand the relative role of genetic polymorphism in genes encoding the variable regions themselves and other genetic and environmental influences on the establishment of the expressed repertoire.

REFERENCES
1. Klinman, N.R., and Press, J.L. (1975). Transplant. Rev. 24:41.
2. Kreth, H.W., and Williamson, A.R. (1973). Eur. J. Immunol. 3:141.
3. Kohler, G. (1976). Eur. J. Immunol. 6:340.
4. Cancro, M.P., Gerhard, W., and Klinman, N.R. (1978). J. Exp. Med. 147:776.
5. Sigal, N.H., and Klinman, N.R. (1978). Adv. in Immunol. Acad. Press. 26:255.
6. Imanishi, T., and Makela, O. (1973). Eur. J. Immunol. 3: 323.
7. Blomberg, B., Geckeler, W., and Weigert, M. (1972). Science 177: 178.
8. Cosenza, H., and Köhler, H. (1972). Science 176: 1027.
9. Lieberman, R., Potter, M., Mushinski, E., Humphrey, W., and Rudikoff, S. (1974). J. Exp. Med. 139: 983.

10. Cohn, M., Blomberg, B., Geckeler, W., Rashke, W., Riblet, R., and Weigert, M. (1974). In: The Immune System: Genes, Receptors, Signals. E. E. Sercarz, A. R. Williamson, and C. F. Fox, eds. Academic Press, New York, p. 89.
11. Sigal, N. H., Cancro, M. P., and Klinman, N.R. (1977). In: Regulation of the Immune System: Genes and the Cells in which They Function. E. E. Sercarz and L. A. Herzenberg, eds. Academic Press, New York. p. 217.
12. Seidman, J. G., Leder, A., Edgell. M. H., Polsky, F., Tilgham, S. M., Tiermeier, D. C. and Leder, P. (1978). Proc. Nat. Acad. Sci. U.S.A. 73: 203.
13. Hood. L. (This volume).
14. Tonegawa, S. (1976). Proc. Nat. Acad. Sci. U.S.A. 73: 203.
15. Wall. R. (This volume).
16. Klinman, N. R., Press, J. L., Sigal, N. H., and Gearhart, P.J. (1976). In: The Generation of Antibody Diversity: A New Look. A. J. Cunningham, ed. Academic Press, p. 127.
17. Klinman, N. R., Sigal, N. H., Metcalf, E. S., Pierce, S. K. and Gearhart, P. J. (1977). Cold Spring Harbor Symp. Quant. Biol. 41: 165.
18. Kabat, E. A., Wu, T. T., and Bilofsky, H. (1978). Proc. Nat. Acad. Sci. U. S. A. 75: 2429.
19. Capra, J. D., and Kindt, T. J. (1975). Immunogenetics. 1: 417.
20. Gally, J., and Edelman, G. (1970). Nature (Lond.) 227: 341.
21. Klinman, N. R., and Press, J. L. (1975). J. Exp. Med. 141: 1133.
22. Klinman, N. R., and Press, J. L. (1975). Fed. Proc., Fed. Am. Soc. Exp. Biol. 34: 47.
23. Sigal, N. H., Gearhart, P. J., Press, J. L., and Klinman, N. R. (1976). Nature (Lond.) 259: 51.
24. Sigal, N. H. (1977). J. Immunol. 119: 1129.
25. Gerhard, W., Braciale, T., and Klinman, N. R. (1975). Eur. J. Immunol. 5: 720.
26. Gerhard, W. (1976). J. Exp. Med. 144: 985.
27. Cancro, M. P., Wylie, D. E., Gerhard, W., and Klinman, N. R. (1979). Proc. Nat. Acad. Sci. U.S.A. (In Press).
28. Pierce, S. K., Metcalf, E. S., and Klinman, N. R. (1979). In: Cells of Immunoglobulin Synthesis. Acad. Press. p. 253.
29. Gershon, R. K. (1976). In: The Generation of Antibody Diversity: A New Look, A. J. Cunningham, ed. Academic Press, p. 105.
30. Cunningham, A. J. (1976). In: The Generation of Antibody Diversity: A New Look, A. J. Cunningham, ed. Academic Press, p. 89.
31. Laskin, J. A., Gray, A., Nisonoff, A., Klinman, N.R., and Gottlieb, P. D. (1977). Proc. Nat. Acad. Sci. U.S.A. 74: 4600.
32. Mäkelä, O., and Karjalainen, K. (1977). Transplant. Rev. 34: 119.
33. Nisonoff, A., and Bangasser, S. A. (1975). Transplant. Rev. 27: 100.
34. Eichmann, K. (1974). Eur. J. Immunol. 4: 296.
35. Cancro, M. P., Sigal, N. H., and Klinman, N. R. (1978). J. Exp. Med. 147:1.
36. Gearhart, P. J., and Cebra, J. J. (1978). Nature. In press.
37. Bell, C., and Klinman, N. R. (Manuscript in preparation).
38. Gearhart, P. J., Sigal, N. H., and Klinman, N. R. (1977). J. Exp. Med. 145: 876.
39. Stashenko, P., and Klinman, N. R. (Submitted for Publication).
40. Karjalainen, K., Bang, B., and Makela, O. (Submitted for Publication).
41. Triplett, E. L. (1962). J. Immunol. 89: 505.
42. Metcalf, E. S., and Klinman, N. R. (1976). J. Exp, Med. 143: 1327.
43. Cambier, J. C., Kettman, J. R., Vitetta, E. S., and Uhr, J. W. (1976). J. Exp. Med. 144: 293.

44. Stocker, J. W. (1977). Immunology 32: 282.
45. Teale, J. M., Layton, J. E., and Nossal, G. J. V. (1979). J. Exp. Med. 150: 205.
46. Metcalf, E. S., and Klinman, N. R. (1977). J. Immunol. 118: 2111.
47. Metcalf, E. S., Sigal, N. H., and Klinman, N. R. (1977). J. Exp. Med. 145: 1382.
48. Cosenza, H., Augustin, A. A., and Julius, M. H. (1977). Cold Spring Harbor Symp. Quant. Biol. 41: 709.
49. Pawlak, L. L., Hart, D. A., and Nisonoff, A. (1973). J. Exp. Med. 137: 1442.
50. Eichmann, K. (1974) J. Immunol. 4: 296.
51. Nisonoff, A., Ju, S-T., and Owen, F. L. (1977). Transplant Rev. 34: 89.
52. Accolla, R. S., Gearhart, P. J., Sigal, N. H., Cancro, M. P., and Klinman, N. R. (1977). Eur. J. Immunol. 7: 876.
53. Rodkey, L. S. (1974). J. Exp. Med. 139: 712.
54. Jerne, N. K. (1974). Ann. Immunol. (Inst. Pasteur). 125: 373.
55. Pierce, S. K., and Klinman, N. R. (1977). J. Exp. Med. 146: 509.
56. Benacerraf, B., and H. O. McDevitt. (1972). Science (Wash. D.C.) 158:1571.
57. Schwartz, R. H. (1978). Scand. J. Immunol. 7:3.
58. Pierce, S. K., Klinman, N. R., Maurer, P. H. and Merryman, C. F. (1979). (Manuscript in preparation).
59. Wysocki, L. J., and Sato, V. L. (1978). Proc. Nat. Acad. Sci. U.S.A. 75: 2844.
60. Mage, M. G., McHugh, L., and Rothstein, T. (1977). J. Immunol. Methods. 15: 47.
61. Julius, M. H., Masuda, T., and Herzenberg, L. A. (1972). Proc. Nat. Acad. Sci. U.S.A. 69: 1934.

(See DISCUSSION on following page)

DISCUSSION OF DR. KLINMAN'S PRESENTATION

Stavitsky: I'm reminded of the experiments of Arthur Silverstein done
many years ago. Has anything like that programmed sequence of
specificities emerged from your work?

Klinman: The story of programmed sequence of responsiveness goes back
quite far. It includes the work of Silverstein and others such
as Montgomery. I think that one can say that our findings are
entirely consistent with that set of findings. But the older
findings have been extended in a sense, in another way. One
never knows, when looking at an entire response to an antigen,
what is the controlling factor. Here we are looking at
clonotypes and the sequential expression of clonotypes, not
necessarily the responsiveness, and in every situation where
clonotypes have been looked at, reproducibility of expression
has been seen. So, I think it confirms and extends those
original findings.

Battisto: Were you suggesting, Dr. Klinman, that cells that were
selected by your technique at the stem cell level were
tolerizable before they showed surface Ig?

Klinman: Ellen Vitetta alluded to this very important concept. Any
experiment that one looks at is totally subject to the system
one is using to analyze. Our culture system has been chosen
for these kinds of analyses because you can isolate the cells
for certain characteristics and then giving them time to
differentiate in vitro and in vivo over a period of time. We
do not imply, at all, that the stem cells we pull out, that
do have IgM are, in effect, affected at all, but rather that
we give them a chance to express their IgM in the presence

of optimum conditions. The intentions of these types of experiments are quite different from those which are trying to examine the triggering mechanism, per se. I have only intended that once we inject the stem cells we can examine them just as easily as we can examine 15-day fetal liver cells or B cells in the thymus. Any B cell population is equally stimulated in the system, largely because it is given an infinite chance with saturation of helper T cells and essentially no suppression.

Index